NORTH CAROLINA
STATE BOARD OF COMMUNITY COLLEGES
LIBRARIES
ASHEVILLE-BUNCOMBE TECHNICAL COLLEGE

DISCARDED

JUN 2 6 2025

HANDBOOK OF Polyester Molding Compounds and Molding Technology

HANDBOOK OF Polyester Molding Compounds and Molding Technology

RAYMOND W. MEYER

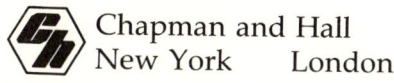
Chapman and Hall
New York   London

First published 1987
by Chapman and Hall
29 West 35 Street, New York, N.Y. 10001

Published in Great Britain by
Chapman and Hall Ltd
11 New Fetter Lane, London EC4P 4EE

©1987 Chapman and Hall

Printed in the United States of America

All Rights Reserved. No part of this book may be reprinted, or reproduced or utilized in any form or by any electronic, mechanical or other means, now known or hereafter invented, including photocopying and recording, or in any information storage or retrieval system, without permission in writing from the publishers.

Library of Congress Cataloging-in-Publication Data

Meyer, Raymond W., 1918–
  Handbook of polyester molding compounds and molding technology.

  Bibliography: p.
  Includes index.
   1. Polyesters. 2. Plastics—Molding. I. Title.
TP1180.P6M49 1986   668.4′12   86-11724
ISBN 0-412-00771-1

Acknowledgments

Many contributors supplied information and photographs for this book. I particularly want to mention the following: Paul Troder of Allied Moulded Products, Robert Dieterle of Amoco, Dave Clavadetcher of Premix, Ivan Brenner of I. G. Brenner Co., Henry Green of Martin Hydraulics, David Evans of Creative Pultrusions, Jim Cobac of BMC Inc., Ed Blue of E. B. Blue Co., Ronald Reid of Chas. Ross & Sons, R. C. Crane of Williams-White, Jai Bajaj of Day Mixing, Bruce Bain of Bonnot, Dean Hermansen of EnTec, Ron Rumpler of OCF, Mike Kallaur of Freeman Chemical, A. H. Horner of Silmar, Bob Talbot and Carl Coats of Ashland Chemicals, Dr. Ray Hindusinn, retired, and Ken Wylegala of Noury Chemical.

Special recognition should go to Nick Mindeola of Gougler for his review and suggestions on the chapter on mold design, to Dick Pistole of Rockwell International, to Dick Savage, retired, and to Nick Bloomer of Haden Research & Development for their help on the painting chapter.

CONTENTS

1	General Information	1
2	FRP Part Design	19
3	FRP Mold Design	47
4	FRP Raw Materials	65
5	Formulations & Compound Preparation	113
6	Molding Compound Process Equipment	147
7	Compression Molding Technology	179
8	Injection Molding Technology	205
9	FRP Painting	221
	Annotated Bibliography	253
	Index	369

HANDBOOK OF
Polyester Molding Compounds and Molding Technology

CHAPTER 1

General Information

Introduction

Initially, glass-reinforced polyester molding compounds were termed *gunks*, and their chief advantage was their ability to mold thick and thin sections in a single part. They were relatively wet, unsophisticated molding compounds. After considerable use and modifications their designation was changed to *premix*, which was defined as an admixture of resin, reinforcement, fillers, etc., not in web form, usually prepared by the molder just before use (ASTM, 1964). The advantages of premix compounds were:

1. the ability to mold thick and thin sections in a part;
2. the ability to mold in metal inserts;
3. excellent electrical properties;
4. lower raw materials costs.

Disadvantages of premix compounds were:

1. poor surface finish;
2. internal cracks and voids in thick cross sections;
3. a tendency for the resin to flow away from the glass reinforcement on long flow paths;
4. relatively low physical and mechanical properties;
5. difficulty in preparing mold charges.

The addition of various thermoplastics offset the polymeric shrinkage that is responsible for the poor surfaces and also eliminated the cracking in thick sections. Those improved molding compounds are today called bulk molding compounds (BMC) in the United States and dough molding compounds (DMC) in Europe. Bulk molding compounds and DMC generally contain 15–20% glass content by weight in lengths of ¼–½ inch. The addition of Group II metal oxides and hydroxides, or chemical thickeners, to increase the compound viscosity helped overcome the problem of the resin flowing away from the glass reinforcement. A further development in this chain of materials evolution resulted in increasing the glass content to approximately 30% by weight and the glass fiber lengths to from 1 to 2 inches to greatly improve the general property level and to simplify the manufacture and handling of the compound by providing it in convenient sheet form. Those latter materials became known as sheet molding compounds (SMC).

It is interesting to note that each of the above changes resulted in a higher viscosity molding compound, which produced more back pressure in the mold during the molding operation.

Description of BMC Product

Bulk molding compounds are completely formulated preimpregnated fiberglass reinforced plastics molding materials that are either prepared by the molder or are purchased from a compounder (see Table 1.1). They can be chemically thickened, in which case generally it is important to control the storage environment to minimize the maturation period or reaction time. When thermoplastic additives are included to control surface finish or surface appearance, the material generally is referred to as low shrink (LS) or low profile (LP) BMC.

Figure 1.1 shows the effect of several formulation ingredients on the shrinkage of molding compounds.

History of BMC Process

The present-day BMC process had a rather ignominious start. It is generally attributed to a technician at the Glastic Corporation. Sometime in 1949 he was assigned the task of finding

General Information 3

Table 1.1 Commercially Available BMC Compounds From BMC Inc.

Series	100	200	300	400	500	700	800	900	1000
Type Compound	General-Purpose Low-Cost	Elec. Gr. Medium-Strength	General-Purpose Low-Shrink	Elec.-Gr. High-Str. Low-Shrink	General-Purpose High Heat Dist.	Elec.-Gr. High Arc Quench	Corr. Res. High-Str.	Encapsulation Soft-Flow	Food Grade
Specific Gravity	1.95	1.95	1.90	1.88	2.0	1.85	1.88	1.90	1.88
Barcol Hardness	40–50	40–50	30–40	30–40	40–50	35–40	40–50	30–35	20–30
Water Absorption	0.07	0.08	0.12	0.14	0.06	0.07	0.05	0.08	0.09
Flexural Strength, psi	13–17,000	10–15,000	18–20,000	18–20,000	13–15,000	10–12,000	18–20,000	9–10,000	16–18,000
Tensile Strength, psi	5–7,000	4–6,000	6,000	6,000	5,000	3–5,000	6,000	4–5,000	5–6,000
Impact Str., Ft. lbs/In N.	3–4	3–4	7–9	6–8	3–4	2–3	5–6	3–4	3–4
Heat Distortion, °F.	500	500	500	500	500	400	500	500	500
Dielec. Str., VPM	430	420	400	350–400	430	350	400	400	N/A
Arc Resistance, Sec.	180	200	180	200	180	200	200	245	N/A
Mold Shrinkage, Mil./In.	2–4	2–4	0–2	0–1	2–3	3–4	1–2	0–1	1–2
U. L. Listed	Yes	Yes	No	Yes	No	Yes	No	Yes	No

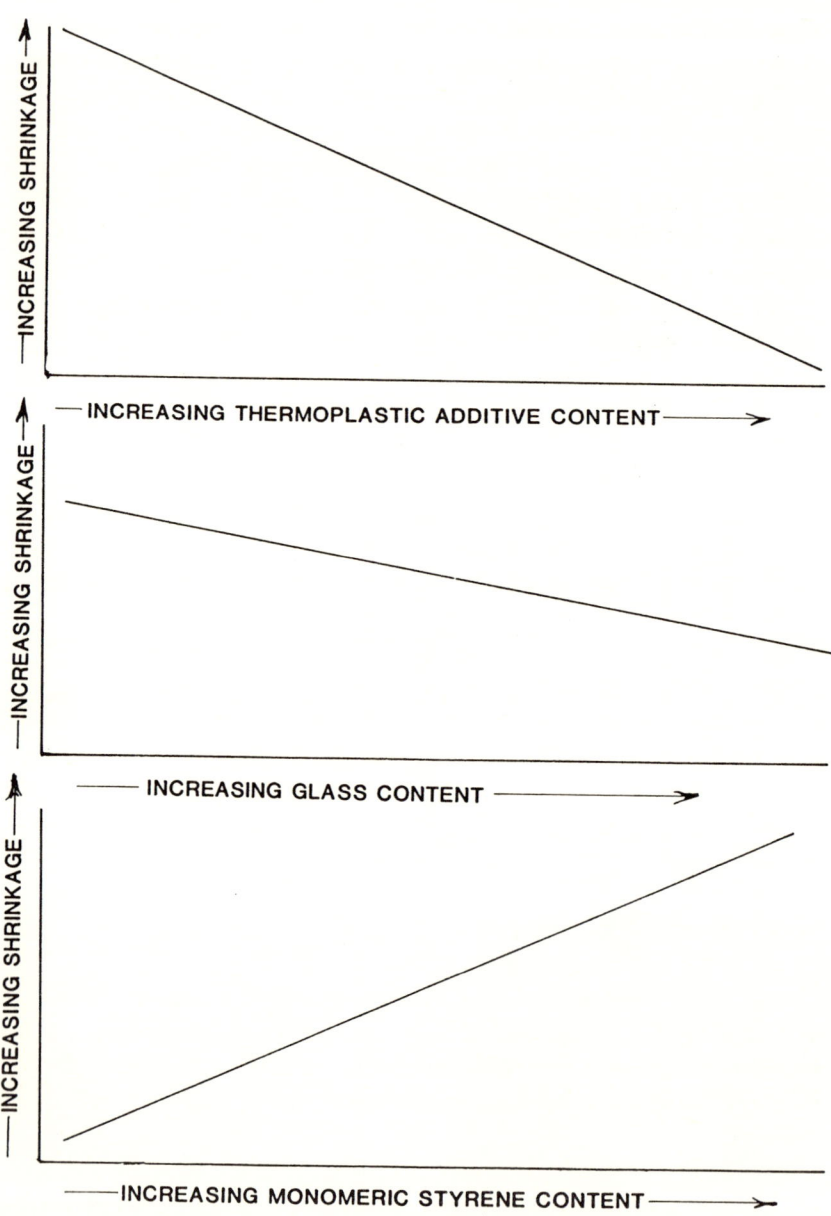

Figure 1.1   General Trends of Shrinkage as Formulations Vary

a use for the mountains of mat trim scrap left over on compression molded parts reinforced with tailored mat. The technician cut the mat trimmings into small squares and hand mixed them into a catalyzed resin for use as a localized reinforcement.[19] The term *gunk* was easily associated with such a mixture. It was soon discovered that the molding material was much easier to handle when the mixture was dried up a little bit by the addition of an inert filler. Soon electrical insulators became a natural product for those materials, and the molding materials became known as *premix*.

Description of SMC Product

Sheet molding compounds are completely formulated preimpregnated, chemically thickened fiberglass reinforced plastics molding materials that are supplied to the press in a dry or nontacky sheet form normally sandwiched between 2 carrier films. The carrier films are removed and discarded immediately before molding. Table 1.2 contains a listing of one popular grade of commercially available SMC together with their physical properties at several glass concentrations. Figure 1.2 consists of a plot of glass content versus physical properties for Premi-Glas™ 7200 SMC.

History of SMC Process

The unique feature of sheet molding compounds is that they are chemically thickened. Early background work in chemical thickening of unsaturated polyester resins was accomplished by investigators who had no interest in molding compounds as such.

In 1951 Weaver[6] described using Group II metal oxides, hydroxides, or carbonates to thicken unsaturated polyester resins as a means of improving the water resistance and electrical

---

Superscript numbers refer to the Annotated Bibliography.

Table 1.2 Typical Physical Properties of Commercially Available Molding Compounds

Glass Content, Percent by Weight	15	22	30	Test Method
	Premi-Glas® 2100-CR-SX			
Specific Gravity	1.80	1.82	1.85	
Flexural Strength, psi	13,000–15,000	14,000–16,000	16,000–17,000	D-790
Tensile Strength, psi	4,000–5,000	5,000–6,000	6,000–7,000	D-638
Izod Impact Strength, Ft./Lbs/In. N.	3–5	4–6	6–7	D-256
Heat Distortion Temp, °F	400+	400+	400+	D-648
Shrinkage, In/In	0.001	0.001	0.001	
	Premi-Glas® 7200 SMC			
Specific Gravity	1.77	1.80	1.85	
Flexural Strength, psi	15,000–17,000	20,000–22,000	25,000–30,000	D-790
Tensile Strength, psi	6,000–7,000	7,000–9,000	10,000–14,000	D-638
Izod Impact Str., Ft.lb/In. N.	6–7	9–10	11–13	D-256
Heat Distortion Temp, °F	400+	400+	400+	D-648
Shrinkage, In/In	0.0002	0.0002	0.0002	

NOTE: Premi-Glas is a Registered Trade Mark of Premix, Inc.
Source: SMC/BMC; *State of the Art*, published by Premix, Inc. Reprinted by permission.

properties of resin castings and adhesives. That same year Frilette[7] was granted a United States patent for the use of Group II metal oxides to harden polyester resins without the use of external heat. The most important work was performed by Fisk[10], who received a United States patent in 1953 for using magnesium oxide to increase the viscosity of unsaturated polyester resins.

The first molding materials to use the thickening principle were premixes. In 1962, at the 17th SPI Reinforced Plastics Division conference, a paper by Cutler, Ferriday, and Parker[75] mentioned crack-free molded parts of an improved DMC or premix.

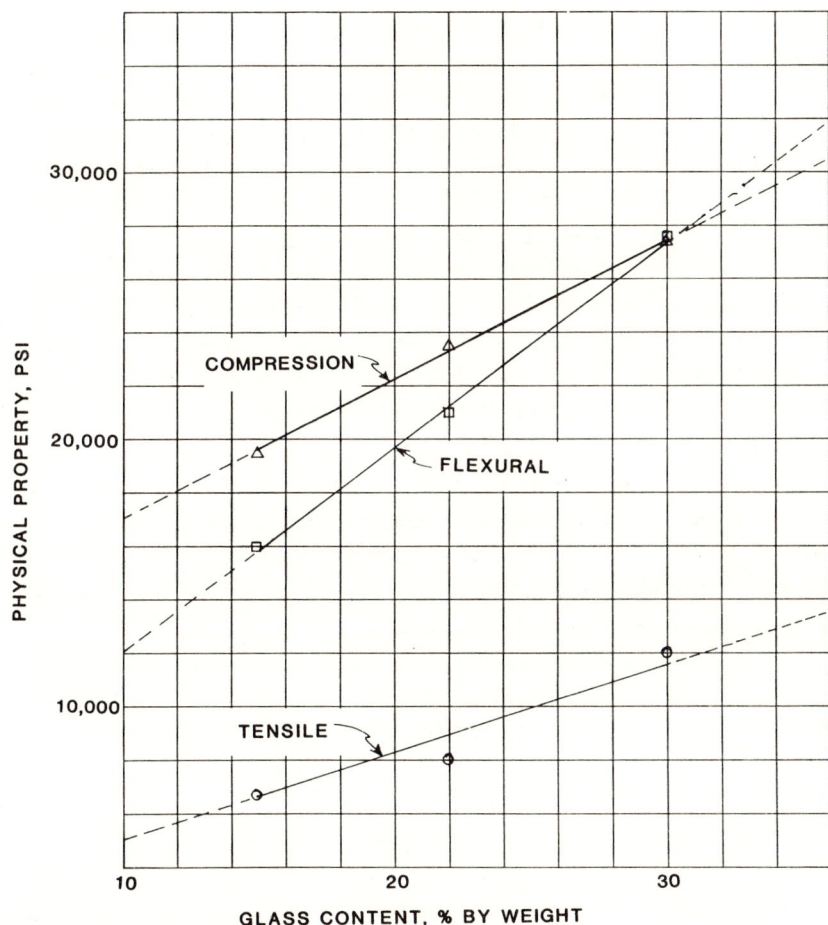

Figure 1.2 A Plot of Physical Properties Vs. Glass Content for Premi-Glas™ 7200 SMC

It is believed that those compounds contained both chemical thickeners and thermoplastic additives. Later that same year a paper by Scholtis[77] described prepreg materials, their properties, and processing methods. The prepregs were reinforced with chopped strand mats.

Early in 1965, Norman Eastwood[101] of Fathergill & Harvey Ltd., reported on Flomat. He mentioned that the product had been used commercially for three years. That same year

Schnitzler[104] published a paper describing formulations, manufacturing procedures, molding notes, and mold design suggestions for a preimpregnated molding material that was called prepreg. Later in 1965 both U.S. Rubber Company and Sohio announced their versions of preimpregnated mat-reinforced molding materials. The U.S. Rubber material called Vibrinmat™ was based on previous work by Farbenfabriken Bayer (1962). Their first impregnator was made by the E. B. Blue Company and is reported to be based on a design used as early as 1962 in West Germany. The Sohio material was called Structoform™ and was based on an entirely different principle involving only dry materials and no chemical thickening.

In 1966 more than a dozen technical papers were published as those new materials received wider acceptance. Most early applications were in the electrical and industrial areas, such as trays, toteboxes, electrical housings, tool cases, transformer cases, fluorescent fixtures, phone boxes, machine housings and guards, ski boots, etc., where a surface finish of 300–400 microinches was satisfactory. The materials continued to be referred to variously as preimpregnated molding materials, mold mat, prepreg, etc. Some confusion existed since there already was a prepreg material available involving coated fabrics. At the April 15, 1966, meeting of the SPI Preform and Mat Die Molding Committee in St. Louis an ad hoc committee was formed to recommend a formal name for those new materials in the United States. The term *sheet molding compound* was adopted, and generally it is contracted to SMC. In Europe the terms *prepreg* and *mold mat* continue to be used, especially in Germany.

Early in 1967, the Marco Chemical Division of W. R. Grace Company bought the U.S. Rubber Vibrinmat process and impregnator and moved it to Linden, New Jersey. Sohio also sold its process. Later W. R. Grace sold its polyester resin business to U.S. Steel, which continues to operate as one of the major resin producers. The Sohio process has been discontinued. Also in 1967, roving was chopped directly onto a moving belt as the input reinforcement rather than using chopped strand mat. That procedure greatly improved the SMC product, because:

1. the unbonded glass fibers flow better;
2. glass-resin ratios are easier to control;
3. various glass lengths can be employed at the same time;
4. cost reductions are possible.

Later a patent was issued on that belt type process (Davis, Wood, and Miller, 1971).[235]

In 1969 low-profile additives became popular, and the first exterior production automotive part was approved. Transportation applications account for the largest volume of SMC in the United States today.

Low Profile Discussion

Low profile or *low shrink* are terms that are applied to BMC and SMC systems that contain a thermoplastic additive which, among other things, counteracts the polymerization and thermal shrinkages in the molded part as the system cures. The term *low profile* is taken from the relatively shallow peaks and valleys on the profilometer measurements made on the molded part surface. The earliest published references to the term *low profile* were in 1967.

All unsaturated polyester resins shrink (5–9%) depending upon the resin composition and styrene concentration. The judicious use of high levels of inert fillers minimizes the shrinkage. To control or offset this shrinkage, a low profile additive or thermoplastic resin is included in the formulation. The low profile additive must become immiscible with the polyester resin at no later time than the start of the curing process. In order for this phase separation to take place, high temperature and pressures are necessary.

Walker[231] provides a good account of the mechanisms involved in low profile systems and lists the following observations as being typical of the systems:

1. The normal gross shrinkage of a part is reduced, eliminated, or converted to a net expansion.
2. Short-range roughness owing to "fiber pattern" is reduced or eliminated so that the molded part has an exceptionally smooth surface.
3. The (resin-thermoplastic) system becomes opaque at some point during the cure cycle.
4. "Hot" (reactive) polyester resins are required.
   a. The number of double bonds per unit molecular weight of polyester chain must be high.
   b. Considerable exotherm is developed during cure.
5. The system contains large amounts of styrene relative to earlier SMC and BMC systems.
6. While there appear to be specific restraints on the nature of the polyester portion of the system, a wide latitude on the nature of the thermoplastic additive is allowable.
7. Materials, such as tertiary butyl catechol, that inhibit homopolymerization (e.g., of styrene) reduce the low-profile effect.

Walker[231] summarizes his article by stating that the low profile resins are complex, multiphase systems that achieve low shrink and smoother surfaces through the interaction of a number of chemical and physical mechanisms. Among them are polymerization shrinkage, thermal expansion and contraction, phase separation, and nucleated foaming under the vapor pressure of residual monomer.

Surface smoothness values of 200 microinches for SMC and 100 microinches for BMC measured on molded flat sheet samples are readily obtainable. On molded parts such surface smoothness values are not easily obtainable and are available only on horizontal areas that involve simple flow patterns. Vertical surfaces do not receive the same molding pressure as do horizontal surfaces, and that factor tends to degrade the molded finish, which becomes more serious as the draw increases.

The specific thermoplastics that can be used to provide the low profile effect will be more fully discussed in chapter 4.

Figure 1.1 shows the general trends in compound shrinkage as the quantity of thermoplastic additive, glass, and styrene monomer is varied in a formulation.

Thickening Mechanism

Sheet molding compound and some bulk molding compounds are chemically thickened. The goal in chemical thickening is to achieve polyester molecules and to be able to control the means of achieving them. Hydroxyl-terminated unsaturated polyester resins may be thickened or forced to undergo a very high viscosity increase by many of the Group II metal oxides and hydroxides including CaO, $Ca(OH)_2$, MgO, $Mg(OH)_2$, and combinations of those materials. The particle size, pretreatment, and dispersion in the matrix mix all influence thickening. A threshold minimum quantity of thickener is necessary to ensure the proper chemical activity. Below the threshold thickener concentration only a minor thickening action takes place. The use of excessive thickener concentration may cause unsatisfactory flow properties and give an unsatisfactory compound. That thickening phenomenon permits a low viscosity resin mix to be used, which allows high inert filler loadings and good wet out of the glass reinforcement to develop maximum physical properties before the controlled chemical thickening or maturation period. Of course, the key term is *controlled*, which is accomplished by judiciously using additives, by maintaining a controlled water content in all raw materials, and by subjecting the completed compound to a slight temperature history. Because of the high filler loading and high viscosity, the compound flows better in the hot mold, resulting in less fiber segregation and better surface finish in the molded part.

The thickening phenomenon has been the subject of considerable research in order to establish the mechanisms involved.

Vansco[206] studied the mechanisms of chemical thickening and proposed that two reactions take place in sequence:

1. salt formation between the metal compound and the acid end group of the polyester resin;
2. a coordinate bond between the basic salt formed in the first reaction and the −CO groups of the ester molecule.

The basic salt formed in the first reaction is of low molecular weight and would not on its own account for the large viscosity rises observed; hence, the secondary reaction was proposed to explain the higher viscosity. That explanation was accepted for several years, but other researchers continued to study the chemical thickening reaction. Burns[597] proposed that the secondary reaction was the result of the formation of a high-molecular-weight linear polymer from carboxylic acid species in the polyester resin. The latter proposal is now generally accepted.

It has been observed that the following variables affect thickening or the rate of thickening:

1. The type and concentration of the Group II metal compound
2. Temperature (heat history) of the reaction
3. Available free moisture content
4. Type and amount of inert filler
5. Presence of free unreacted glycol or other −OH containing compound
6. Presence of formamide or other amide.

Types of Molding Compounds

Numerous grades of molding compounds are available. In this section some of the more popular types will be described. The raw materials that differentiate the several grades and formulations are discussed in chapter 5.

General Purpose

General purpose molding compounds are one of the more simple and lower-cost molding compounds. They are heavily loaded with inert fillers and contain only enough glass fiber reinforcement to meet a specific end use requirement. Their good flow and ability to mold thick and thin areas within a part and to have inserts molded in are their chief advantages. Their principle disadvantage is high shrinkage of the polyester resin, which results in uneven or relatively rough surface finish. There are several applications where surface finish is unimportant, such as on some industrial products, and the lower cost of those general purpose compounds can be a benefit in such cases.

Electrical Grade

Electrical grade molding compounds are different from general-purpose compounds in the type of inert fillers used and the fact that they are flame retardant. China clays or calcium sulfate are used to improve some electrical properties and water absorption. For arc quenching or tracking resistance alumina trihydrate filler is used. In heavily filler-loaded compounds the addition of small quantities of antimony trioxide makes them flame retardant. Chlorinated or brominated wax compounds can be added to improve the flame retardant rating. For the ultimate compound, a halogenated polyester resin in combination with antimony trioxide is used. Figure 1.3. shows a popular item molded of an electrical grade compound.

Chemically Resistant Grade

A chemically resistant molding compound can be prepared by removing the calcium carbonate from a general-purpose grade molding compound and replacing it with china clay. If flame-retardant properties are also desired, then antimony trioxide and chlorinated paraffin wax can be included. Isophthalic polyester resins are used for the lower end compounds, while vinyl

14   Handbook of Polyester Molding Compounds and Technology

Figure 1.3 Various Premix Electrical Outlet Boxes (*Courtesy: Allied Moulded Products*)

ester and bisphenol-A resins are used where maximum chemical resistance is desired.

Low Profile/Low-Shrink Grade

The terms *low profile* and *low shrink* have not been rigidly defined and are used interchangeably in many cases, but generally *low profile* means Class A automotive surface finish while *low shrink* means a good surface finish but not quite a Class A finish. The differentiation generally has to do with the type and quantity of thermoplastic additive used. Polystyrene and powdered polyethylene are used to produce low shrink finishes. For low profile finishes polymethyl methacrylate, polyvinyl acetate, etc., are used.

Food Grade Compounds

Where products come into contact with foods, a resin approved by the Food and Drug Administration must be used. The FDA is generally specific regarding the type and quantity of organic peroxide, pigments, etc., that can be included in a formulation.

Low Pressure Molding Compounds

In general, most conventional BMC is molded at pressures of from 500 to 1,000 psi while SMC molds in the range of 1,000 to 1,500 psi. Some individual compounds may require higher pressures. By controlling the maturation rate and extent with additives, it is possible to make chemically thickened compounds that will mold at substantially lower pressures. Those additives and the typical formulations are discussed more fully in chapter 5.

Low Density Molding Compounds

The density of conventional grades of BMC and SMC falls in the range of 1.7–1.9 g/cm^3. By the use of syntactic foam fillers to replace a portion of the inert filler, that density can be reduced to a range of 1.2–1.4 g/cm^3.

High Strength Molding Compounds

By reducing or eliminating the inert filler content, by increasing the glass content, and by eliminating the thermoplastic additive, a high-strength structural grade molding compound can be prepared that is satisfactory for many under-the-hood and out-of-the-way automotive applications where appearance properties do not apply.

Energy Absorbing Compounds

"Tough" or energy absorbing molding compounds either use specially formulated resins or include "toughener" additives in the formulation. A short description of these latter additives will be found in chapter 4.

Molding Compound Markets

Table 1.3 is a tabulation of the estimated shipments of molding compounds (BMC and SMC) for the years 1976 and 1983, broken down by market area.

While there has been some movement among market areas for BMC, the total shipments for the two periods are almost identical. That lack of growth also is reflected in the lack of development effort on BMC and on their raw materials; such is not the case for SMC materials.

The total estimated shipments for SMC in 1983 amounts to a 65% increase over the 1976 total. The largest market area for SMC is land transportation, which accounts for 63% of the total SMC shipments. Exterior body panels on the Corvette and Fiero automobiles, truck and van parts, spoilers on sports cars, wind deflectors on truck cabs, etc., are some of the automotive applications.

The next largest market area for SMC is in appliances and business equipment, which accounts for approximately 13% of the total SMC shipments. That is closely followed by the construction area, at approximately 11%, and the electrical and electronic market, at approximately 8%.

Table 1.3 Estimates of Molding Compound Shipments By Markets (in millions of pounds of composite)

Material	BMC		SMC	
Year	1976[a]	1983[b]	1976[a]	1983[b]
Aircraft and Aerospace	—	—	6	—
Appliance and Bus. Equip.	8	10	42	37
Construction	24	5	—	30
Corrosion-Resistant	3	10	2	2
Electrical and Electronic	28	31	12	23
Marine	—	—	—	—
Land Transportation	44	35	103	177
Consumer	—	21	3	9
Miscellaneous	6	7	3	4
Total	113	119	171	282
Source	1	2	1	2

[a]SPI RP/CI; Includes both thermoplastic and thermoset composites.
[b]Unofficial estimates.

CHAPTER 2

FRP Part Design

FRP Mechanical Property Design Criteria

The rules for designing with metals, wood, and concrete have been available for more than a century, but the same type of reliable design data has been readily available on fiberglass reinforced plastic (FRP) laminates only in recent years. There is still a need for such data to be generated.

The principles used in structural design with FRP can be compared to designing with steel or aluminum. In the normal design of structures using metals within the elastic range the assumption is made that the metals are never stressed beyond the proportional limit. Within this elastic range the materials are said to obey Hook's law, that the unit stress is directly proportional to the unit strain. No material is perfectly elastic, but it has been found by experiment that most materials used in engineering approximate this condition closely within certain limits. Figure 2.1 contains representative stress-strain curves for boiler-plate steel, cast iron, concrete, timber, and FRP. With FRP materials there is no yield point in the stress-strain curve, as there is with steel. The strain is directly proportional to the applied load up to the point of catastrophic failure.

It is important to consider design factors in designing with FRP laminates in load-bearing structures. The design factor or factor of safety is defined as the ratio of the ultimate strength of a material to the allowable working stress. In many fields the allowable working stresses are specified by a code authority. Most FRP molders can reliably specify the ultimate strengths

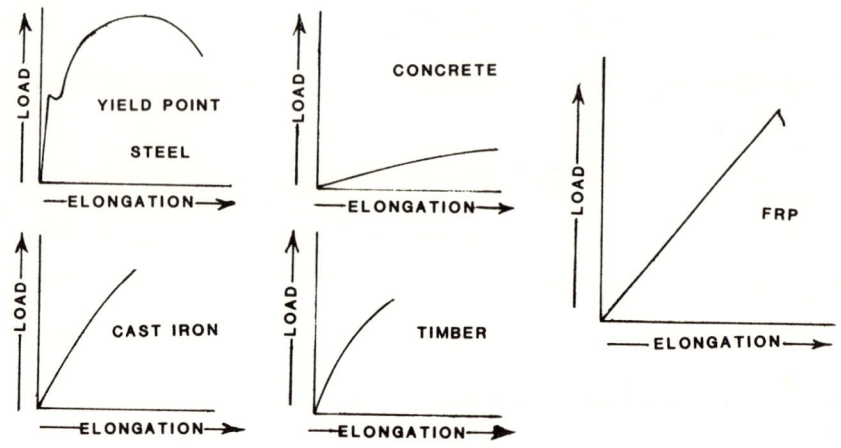

Figure 2.1   Typical Stress–Strain Curves

of their laminates in tension, compression, and flexure, but generally it is up to the part designer to pick his design factors. The nature of the type of loading the structure will undergo and the environmental temperature has a bearing on the design load. The strength properties of FRP tend to increase as the temperature decreases and to decrease as the temperature increases. Table 2.1 lists minimum design factors for several types of loading according to Gaylord[329], and Table 2.2 lists the effect of temperature on one FRP system. If a structure is to be

Table 2.1   The Effect of Type of Loading On Design Factors

Type of Loading	Minimum Design Factor
Static Short-Term Loads	2
Static Long-Term Loads	4
Variable or Changing Loads	4
Repeated Loads, Load Reversal, Fatigue Loads	6
Impact Loads	10

M. W. Gaylord, *Reinforced Plastics—Theory and Practice*, 2nd Edition, Cahners books, New York (1974).

Table 2.2 The Effect of Temperature on One FRP System

Temperature	Tensile Strength	Modulus of Elasticity
75°F	20,000 psi	2.3×10^6
125°F	16,000 psi	1.8×10^6
175°F	12,000 psi	1.4×10^6
200°F	10,000 psi	1.2×10^6

Reference: M. W. Gaylord, *Reinforced Plastics—Theory and Practice*, 2nd Edition, Cahners Books, New York (1974). Reprinted with permission.

subjected to other than normal ambient temperature, it is desirable to test the laminate at that service temperature. Table 2.3 lists the comparative properties of several FRP materials, aluminum, stainless steel, wood, and low carbon steel.

FRP Part Design Principles

The following basic principles of part design will greatly assist in obtaining the best molding behavior from BMC and SMC.

Locate Flash So That It Can Be Easily Removed

On simple parts the designer has the choice of whether to locate the mold parting line on the vertical plane or on the horizontal plane of the part as it is molded. So far as production of the part is concerned, it is best to locate the parting line so as to make removal of the flash as simple as possible. The glass reinforcing fibers toughen the flash as well as the molding compound. In the case of a dish-type part, those two design possibilities are shown in Figure 2.2. A jitterbug sander probably would be used in either case for flash removal. Removing the flash is fairly simple and straightforward when a vertical flash is used, as in the left-hand part of Figure 2.2. The part would be supported in a jig so that the flash extended upward and the operator would move the sander in a horizontal plane over the flash. In regard to the right-hand section of Figure 2.2, the

Table 2.3 Comparative Properties of Several Materials

Property	Pultruded FRP Rod	Press-Molded BMC	Press-Molded SMC	Aluminum Sheet	Stainless Steel	Low Carbon Steel
Glass Content (percent by weight)	70	10–25	25–35	—	—	—
Flexural Strength (psi)	100M	10M–25M	25M–40M	20M	30M–35M	28M
Flexural Modulus (psi $\times$ 10^5)	60	8–12	10–12	100	280	300
Tensile Strength (psi)	100M	4M–8M	12M–20M	6M–27M	30M–35M	29M–33M
Tensile Modulus (psi $\times$ 10^5)	60	5–15	9–20	100	280	300
Impact Strength (ft.lbs/in.)	49	3–10	10–20	—	8.5–11	—
Thermal Conductivity (BTU/hr/ft²/°F/in.)	5	1.4–1.9	1.3–1.8	810–1620	96–185	260–460
Specific Heat (BTU/lb./°F)	0.24	.31–.34	.30–.33	.22–.23	.12	.10–.11
Rockwell Hardness (Scale)	80H	40–80H	40–105H	1–5B	90B	72
Dielectric Strength (volts/mil)	400	300–500	300–600	—	—	—
Specific Gravity (25°C)	2.10	1.8–2.0	1.5–1.7	2.6–2.8	7.92	7.9
Density (lb/in.³)	0.072		.054–.061	.10	.29	.29
Thermal Coeff. of Exp. (in./in.°F $\times$ 10^6)	3	8–15	10–18	12–13	9–10	6–8

FRP Part Design   23

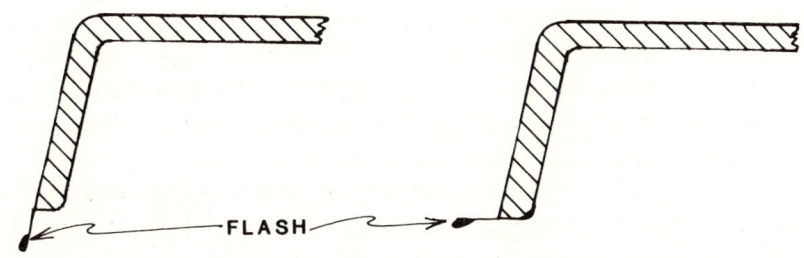

Figure 2.2   Parting Line Possibilities on a Dish-Type Part

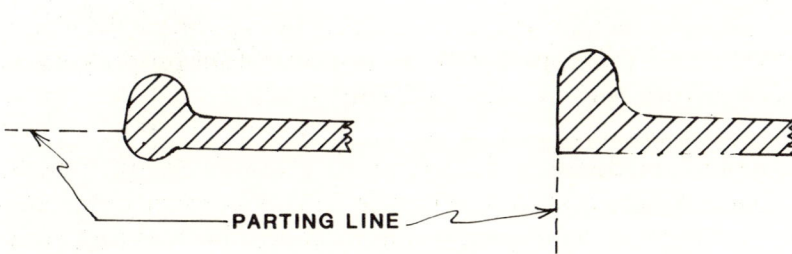

Figure 2.3   Parting Line Possibilities on a Tray-Type Part

part would have to be rotated, and there is danger that the vertical walls near the flash would be gouged with the sander.

Avoid Parting Lines on Show Surfaces

Figure 2.3A shows the cross section of the edge of a rectangular hospital tray made of molded-in color SMC. Both top and bottom edges of the tray were specified to be rounded. In order to accomplish that, the mold parting line had to be put on the horizontal center line of the part, as shown in the left-hand diagram. Sanding off the flash left an unsightly white line all around the tray. That operation resulted in a large quantity of rejected trays, which had to be salvaged by painting, adding extra expense. Redesigning that part according to the sketch in Figure 2.3B would have permitted a vertical parting line. Then

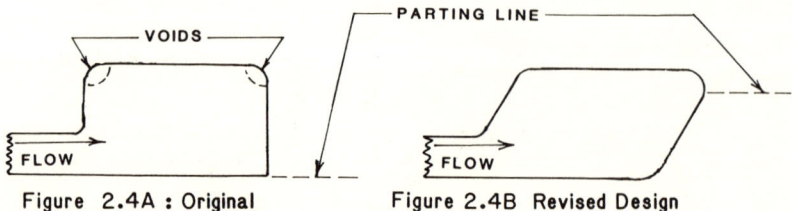

Figure 2.4A : Original   Figure 2.4B Revised Design

Figure 2.4   Parting Line on a Fan-Type Part

the flash could have been sanded off without interfering with the critical exposed edge surfaces, and the part cost probably would have been considerably reduced.

Avoid Blind Pockets

Blind pockets usually occur over deep bosses and over some ribs. When air is trapped in those pockets, it is compressed as the compound fills the part, causing voids. In extreme cases diesel burn results. Such a condition existed in the original design of the blades on a molded fan used to cool electric motors. That blade design is shown in Figure 2.4A. When the fan blades were redesigned, as shown in Figure 2.4B, the molding compound was able to flow into all areas without trapping air. Knockout pins can be placed on bosses and over ribs to act as vents for bleeding entrapped air. Figure 2.5 shows the design of one type of self-cleaning pin that has been used successfully with both BMC and SMC.

Provide a Generous Draft on Vertical Walls, Bosses, and Ribs

Since polyester molding compounds shrink as they cure, it is necessary to provide a draft in order to get a part to release from the mold. On parts with depths of up to 3 inches, a minimum draft of 1° is recommended. On parts from 3 to 6 inches deep a minimum draft of 2° is necessary while a 3° draft is necessary for parts deeper than 6 inches. That is shown diagrammatically in Figure 2.6.

FRP Part Design 25

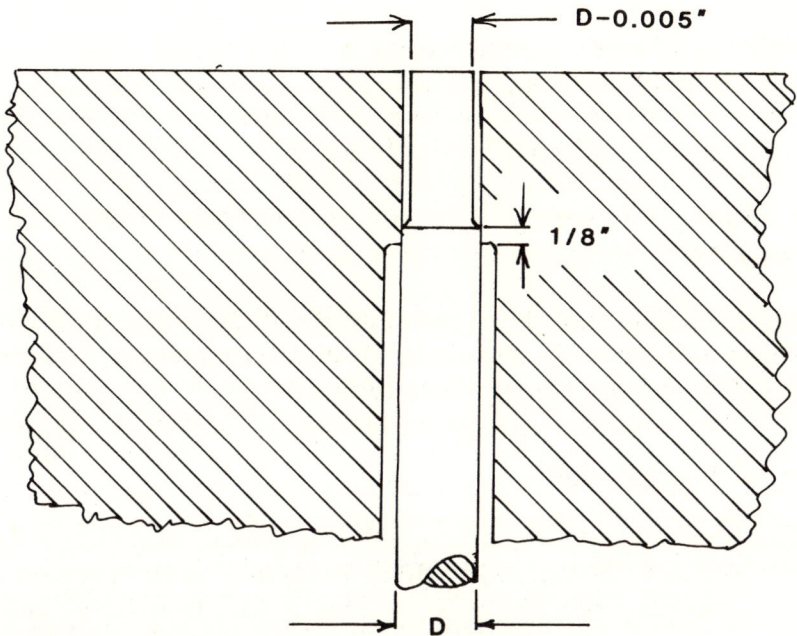

Figure 2.5 Self-Cleaning Knockout Pin Design

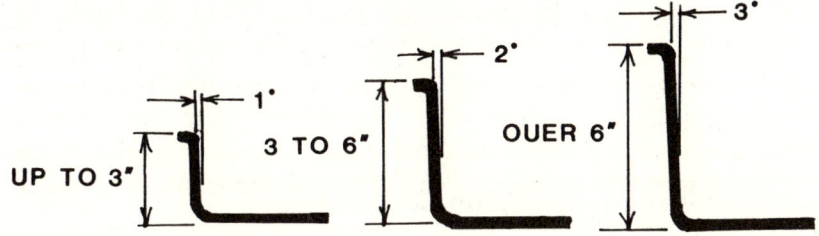

Figure 2.6 Draft Details

Avoid Molded-through Holes When Maximum Strength Is Required

Whenever polyester molding compounds are forced to flow around a core pin in a mold, a knit line results that is weaker than the surrounding areas of the part. Cracks frequently occur at that knit line. For maximum strength it is advisable to drill holes in a secondary operation rather than to mold them in, especially those holes near the outer periphery of the part. If molded-in holes are required for some particular reason, then it is advisable to place the charge over the hole areas and to force the compound to flow away from the holes in all directions and hereby avoid knit lines in those areas.

Avoid Press Outs and Mash Outs Whenever Possible

Including press outs and mash outs in a part design is common practice in preform and mat-reinforced parts since the reinforcement is already in place when the mold closes, and the thin liquid mix still has an opportunity to saturate the reinforcement. However, in BMC and SMC molding the reinforcement must flow with the resin and filler. Placing a press out area in a mold acts as an obstruction to the advancing fronts, which generally must flow around that obstruction, thereby severely orienting the glass reinforcement and weakening the part in at least one direction. It is advisable not to include press-out areas in BMC and SMC part designs. An exception to that principle will be discussed later regarding picture-frame type parts. If press outs or mash outs cannot be avoided in a part design, then the compound charge should be placed over the press out area in the mold to force the flow away from that area to minimize problems.

Avoid Flowing from a Thin Section to a Thicker Section

When the moving front in a mold is allowed to flow from a thin section to a thicker section, the back pressure is lost and the moving front begins to cascade. That action is shown in Figure 2.7. Long suspected by many investigators, it was visually demonstrated by high-speed photography several years ago in a mold constructed by Owens-Corning Fiberglas Corporation, which contained replaceable rib sections and a quartz window through which material behavior could be observed.

Also, several years ago problems were encountered with severe porosity on the outer edges of an automotive fender

FRP Part Design 27

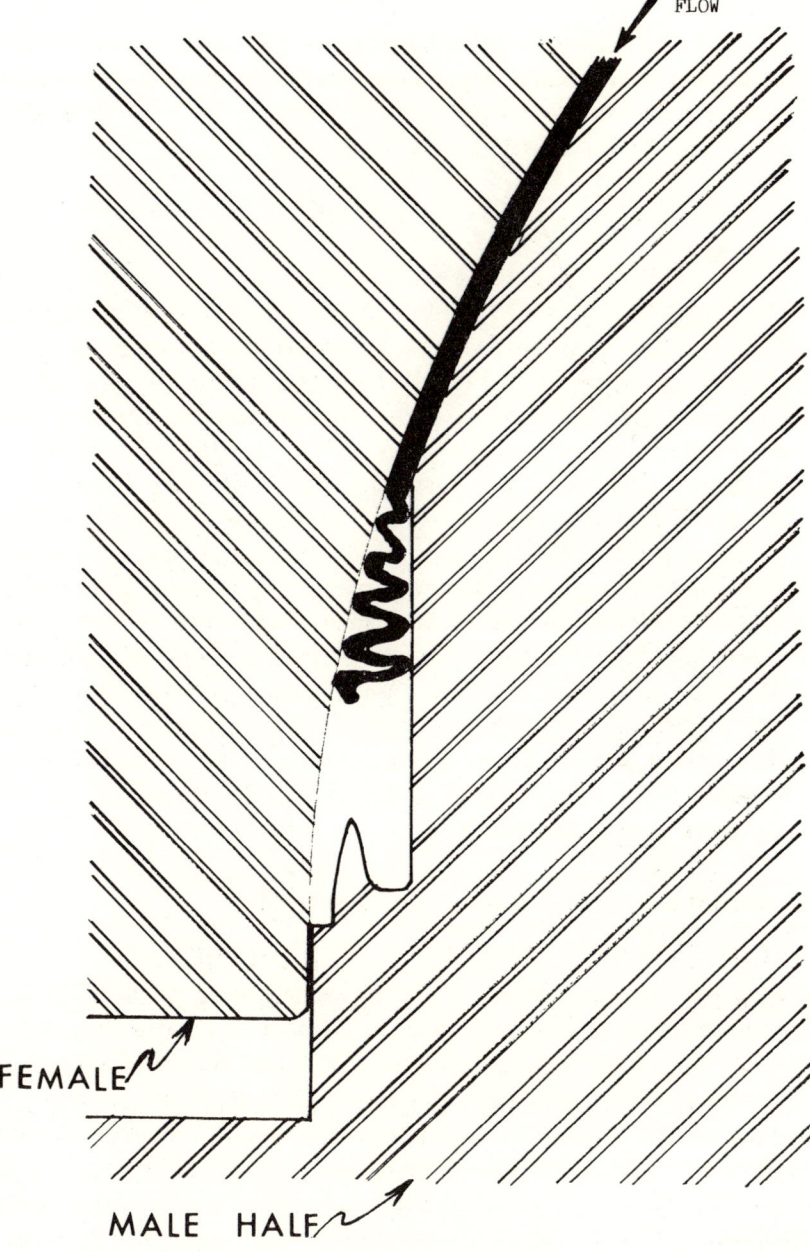

Figure 2.7 Behavior of the Moving Front in a Mold When Flowing from a Thin Section to a Thicker Section

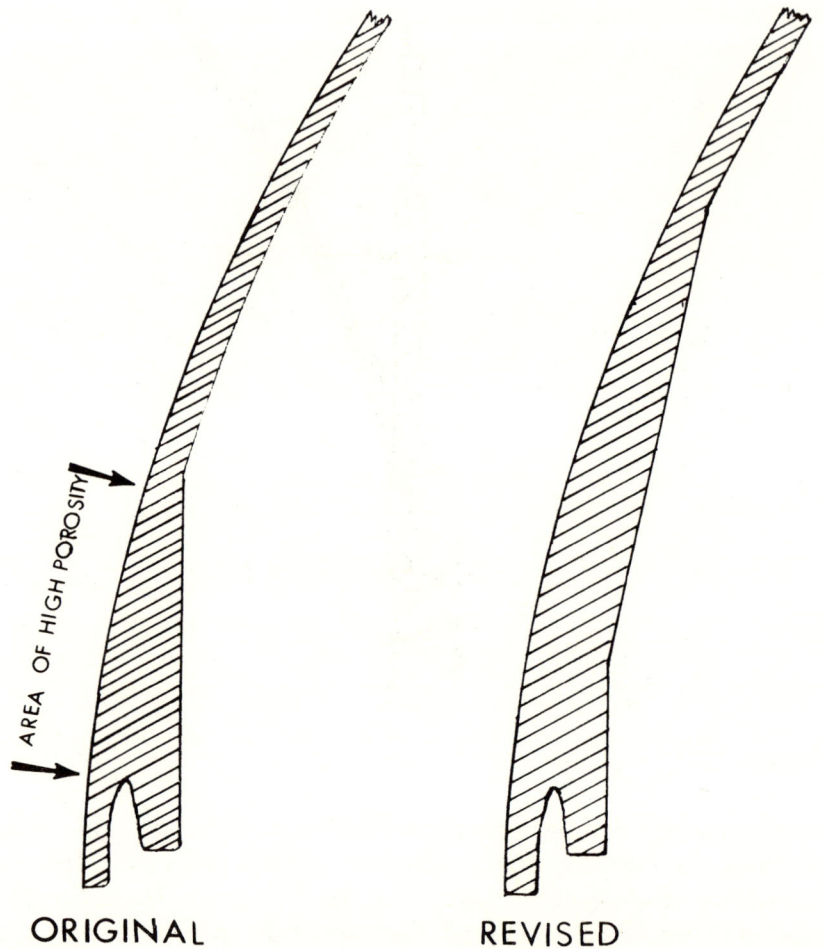

Figure 2.8 Flowing from a Thin Section to a Thicker Section Can Entrap Air, Causing Porosity

extension. A cross section of that outer edge design is shown in Figure 2.8. The original part had been made of die cast zinc. The outer edge contained an indentation to receive a molded rubber gasket to separate the zinc part from the steel fender to eliminate electrolysis and retard body rusting. That part was converted to FRP for a trial on the basis that the FRP part be identical in shape to the original die casting. It was discovered

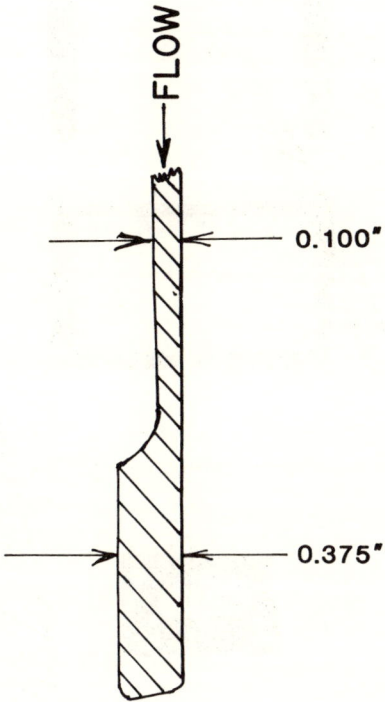

Figure 2.9 An Illustration of Flow from a Thin to a Thick Section

that the area of high porosity coincided with the enlarged edge section. The degree of porosity was greatly reduced when the mold was reworked to minimize the transition from the thin section to the thicker section. The same type of condition existed on a 1975 model SMC truck front end outer edge, which is shown in Figure 2.9. The enlarged section was included as a means of stiffening the entire front end. While that objective was met, an unusual quantity of defects (porosity) resulted.

Molding Picture-Frame Type Parts Presents Problems

Picture-frame-type parts present no unusual problems for preform or mat reinforced parts, but they do for BMC and SMC parts. Both those materials must flow in the mold to develop

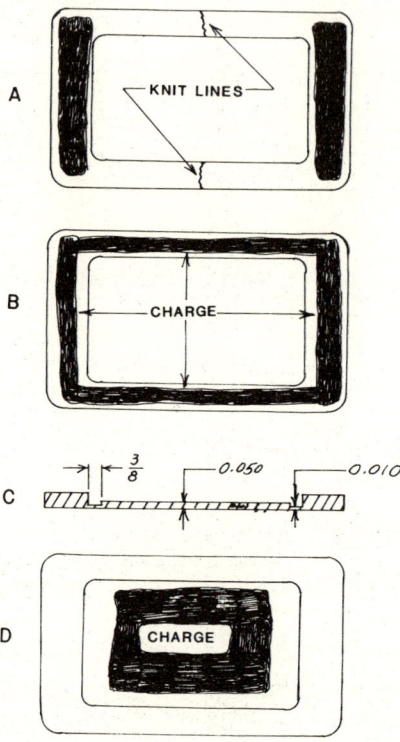

Figure 2.10 Picture-Frame-Type Parts Present Problems

their optimum properties. If a mold is charged, as shown in Figure 2.10A, a weak knit line will develop. Sheet molding compounds can be charged, as shown in Figure 2.10B, but that requires considerable time on the operator's part and precure could result before the mold is fully closed. Four narrow strips of continuous strand mat could be positioned on the mold before adding the charge. The mat would protect the compound from the hot mold and reinforce the knit lines to prevent cracks and weak physical properties. If large quantities of such parts are to be made of SMC, it generally is less expensive to build a press out in the mold, as shown in Figure 2.10C, and to charge the mold, as shown in Figure 2.10D. By that procedure the center section is trimmed in the mold, and the problem with knit lines is eliminated.

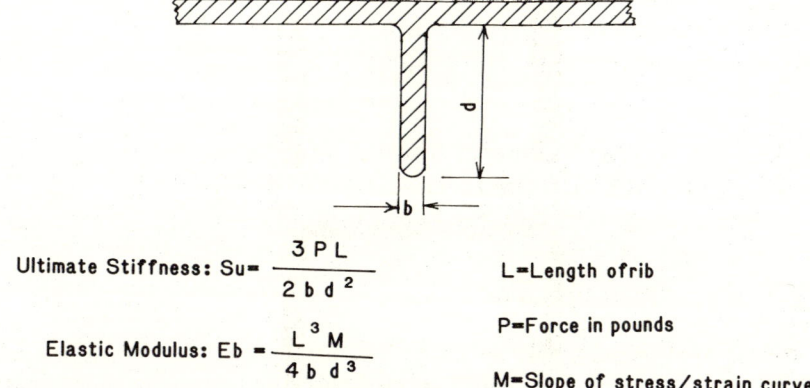

Figure 2.11 Stiffness Calculations for Ribs

Ultimate Stiffness: $S_u = \dfrac{3PL}{2bd^2}$

Elastic Modulus: $E_b = \dfrac{L^3 M}{4bd^3}$

L = Length of rib
P = Force in pounds
M = Slope of stress/strain curve

Avoid Internal Undercuts on a Part

Internal undercuts on an SMC or BMC part must be avoided, but external undercuts may be molded using slides or split molds. Such slides should be located around the periphery of the part. Slides may be operated mechanically or hydraulically.

For trouble-free molding it is wise to follow all of the previously mentioned design principles. That does not mean that parts violating one or more of those principles can not be molded. It does mean, however, that one should expect problems with parts that violate those principles and some allowances should be made in costing for the higher quantity of rejects that can be expected.

Molding Compound Specific Design Details

Ribs on Nonappearance Parts

Use ribs only where they are needed to provide rigidity to a part. The mass of the rib will affect overall cure time of the part so it should be kept to a minimum. Figure 2.11 lists the basic formulae for calculating rib stiffness.

Design considerations are not so crtical for ribs on nonappearance parts as they are for exterior show surfaces. The following general design guides may be used:

1. Ribs should have a minimum taper of 1° each side.
2. Lead-in fillets at the rib bases generally should be 1/16 inches minimum and can be as general as desired.

Ribs on Appearance Parts

Incorporating ribs into appearance parts is a much more complex situation. A first step is proper selection of the molding compound to ensure minimum shrinkage. However, even though minimum shrinkage is to be expected from the compound, the presence of ribs on show surfaces will telescope through unless they are placed under styling lines, contoured surface areas, etc. Figure 2.12 contains a set of guidelines and recommendations for designing ribs into appearance parts.

Bosses

Figure 2.13 consists of a set of guidelines for bosses on both BMC and SMC parts.

Massey[420] reports on a study of the effect of boss type, hole diameter, and glass content on driving torque, torque failure, and pull-out force, using three different self-tapping screws. Englehart[462] published a similiar report.

Inserts

Metal inserts can readily be molded into BMC and SMC parts. Inserts should be located perpendicular to the parting lines and secured so that they will not be displaced during compound flow under pressure. Nickel- and cadmium-plated steel and brass are frequently used. Aluminum has been used on large inserts where keeping weight to a minimum was important. Unplated copper, brass, and zinc inserts should not be used. Copper ions inhibit some polyester resins and prevent complete cure adjacent to the insert. Zinc acts as a promoter and can

FRP Part Design   33

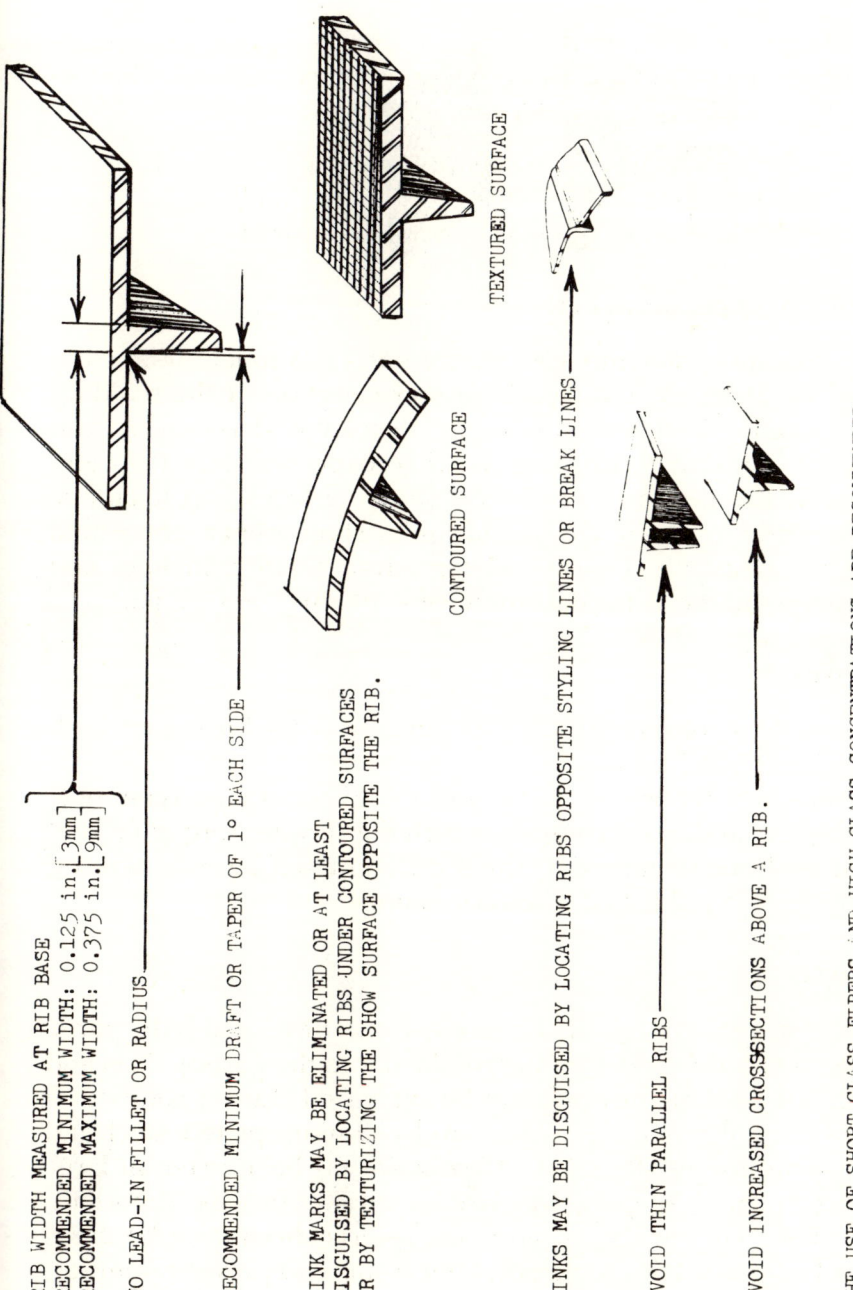

Figure 2.12 Guidelines for Ribs on Appearance Parts

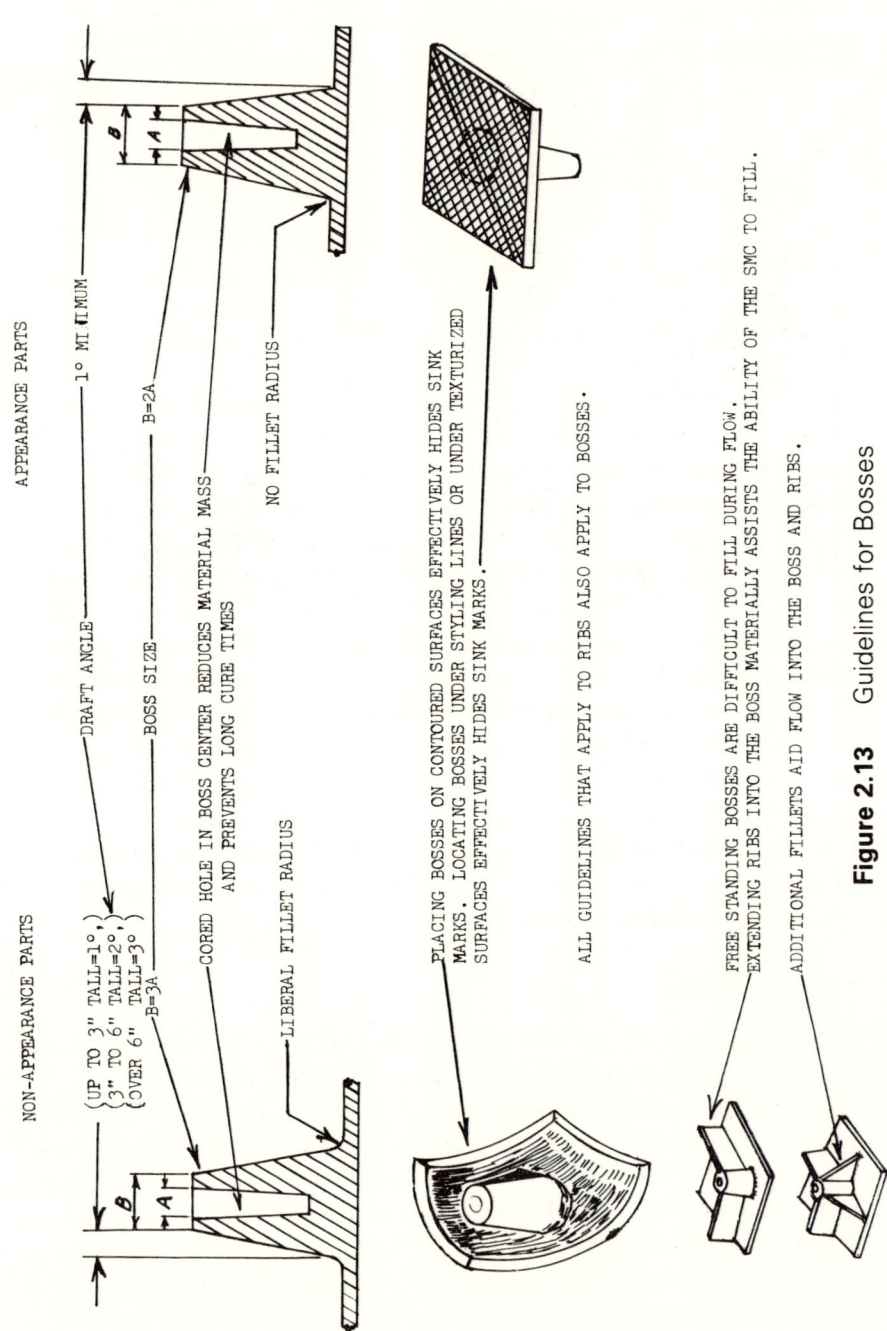

Figure 2.13 Guidelines for Bosses

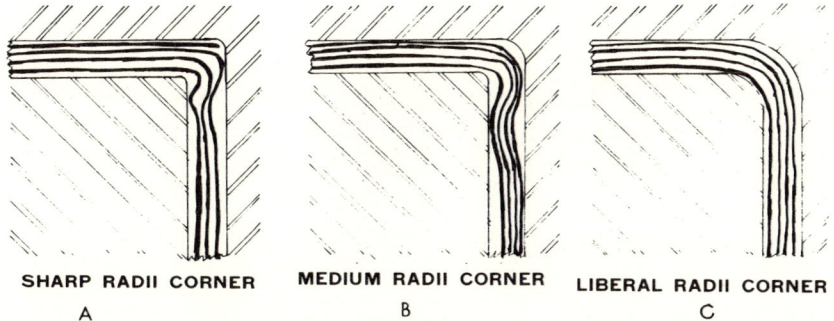

SHARP RADII CORNER	MEDIUM RADII CORNER	LIBERAL RADII CORNER
A	B	C

Figure 2.14 The Effect of Flowing SMC around Corners

precure some resins. Metal inserts can be designed to be either male or female with respect to their attaching hardware. The portion of the insert to be embedded in the molding compound should be cylindrical in shape and contain undercut grooves or deep knurling or both to permit a good bond and good holding power between the molding compound and the metal insert. Square or hexagonal corners on inserts tend to produce areas of high-stress concentration that can crack under load. Metal inserts should be degreased before use and preferably should be at the same temperature as the mold when they are inserted into it. Otherwise the cold insert will act as a heat sink and may interfere with the cure of the compound. That is easily and economically taken care of by storing a supply of metal inserts on the steam platen adjacent to the mold or on corners of the mold, when they are available. The operator must use gloves or a tool to handle the hot metal inserts.

Corner Details

On box type parts where corners are involved, it is important to use the largest corner radius possible. Figure 2.14 shows the effect of glass reinforcement orientation resulting from flowing SMC around various radii corners. Such orientation can cause the walls of thin boxes to warp. That is particularly important if the boxes have been designed to nest with each other or when lids must mate with the box. Proper charge placement also is involved in molding box-type parts. The use of pyramid charges (bottom plies always larger than succeeding top plies) helps avoid weak knit lines on box-type parts. Since BMC generally has much shorter glass lengths, that tendency for glass orientation when flowing around corners is reduced.

Fabrication Techniques

With the introduction of standard FRP structural shapes, it became necessary to develop fabrication techniques for combining FRP members. Several of the larger pultruders have published excellent design manuals covering their products. Most of the fabricating methods used on wood, aluminum, and steel are useful for the fabrication of FRP products. FRP shapes cannot be welded, but they can be adhesively bonded. Some fabricating characteristics of FRP shapes should be considered.

Fiberglass reinforced plastic is extremely abrasive. Tools wear quickly and require sharpening or replacing more often than in working with wood or metals. Maintaining a sufficient inventory of spare tools is important to keep production moving. Tungsten carbide tools are preferred. Diamond-coated edges and saw blades are best. Water cooling adds to the life of diamond tools. Excessive pressure should be avoided during cutting, drilling, and routing.

Machining FRP is a dusty procedure. Minor health problems, such as skin irritations, can be expected on a small percentage of new workers. Fair-skinned people and those with blond or red hair seem to be most affected. Protective ointments are available that help condition the skin of new workers. Masks should be worn in severe dust areas. Long-sleeved shirts and shop coats are recommended. Air-powered hand tools are safest. Totally enclosed electrical motors generally are used, and in the areas of highest dust levels explosion-proof motors should be considered.

Fiberglass reinforced plastic is elastic; warpage and twist can be expected on long lengths. Rigid supports are required for all fabricating and machining operations. Clamps, tie-downs, and vises should be installed on both sides of saw cuts.

Fiberglass reinforced plastics will soften when exposed to heat. Fast-cutting tools generate heat that softens the FRP laminate, causing ragged edges, splitting of the fabricated part, and reduced tool life. Water cooling helps tremendously on sawing and routing operations. Appropriate machine speeds must be used. The greater the FRP thickness, the slower should be the cutter speed on milling and drilling operations.

Drilling

Care must be taken to prevent splitting when drilling parallel to the continuous roving direction (when used) in FRP parts. The FRP part must be clamped in a fixture or vise. When drilling prependicular to the continuous roving direction, that precaution is not absolutely necessary, but it is a good practice to clamp the part and to back it with a metal support to prevent the bit from breaking through the laminate. Specially designed tungsten carbide-tipped drill bits having a positive rake should be used. WONDERtwist™ drills and WONDERdrill™ are two such drills made by International Carbide Corporation. A good discussion of the design and use of those drills is to be found in Mackey's paper.[547] Other such proprietary drills also are available.

Tungsten Carbide hole saws (manufactured by Tunco Manufacturing) are available for drilling larger-diameter holes.

Routing

Router bits should be diamond coated. Some routing operations are done dry, which generates large volumes of fine dust, which must be collected and disposed of properly. Dust in a manufacturing plant is a nuisance to employees and can be an explosion hazard. If water can be used to cool the router bit and prevent it from being clogged with resin particles, the tool life is extended considerably and the dust problem is minimized. The coolant can be collected in a sump and the dust removed by filtration.

Punching

For small holes under ⅜-inch diameter a high-chrome, high-carbon steel is used. For larger holes use a good grade of oil-hardening tool steel. Inserts are machined and heat treated and then ground to size. Clearances of 0.010–0.020 inch are used.

It is recommended that spring-loaded stripper mechanisms be used and that the srings be loaded from 2 to 3 times that normally used for steel punches. There is a tendency for the FRP to cling to the extended punch since the punched hole has a diameter from 0.002 to 0.010 inches smaller than the punch. Holes smaller in diameter than the FRP laminate thickness should not be punched. Punched holes should be kept at least 3 times the laminate thickness from the edge of the part. Punches work better on FRP when they contain a shear angle. A shear angle that is equal to the material thickness will reduce the required press tonnage by one-third. Shear edges also produce cleaner punched holes.

Sawing

Sawing of FRP parts can be done with abrasive wheels, but it is preferable to use a continuous-rim diamond blade that is water cooled. Small-diameter FRP parts can be cut off continuously with a dry diamond wheel, but the improved part appearance, the reduced dust problem, and the extended tool life generally make the extra effort of supplying and collecting the coolant spray advisable and economical.

Milling

Tungsten carbide-tipped cutters are recommended for use at a cutter speed of from 1,500–2,000 surface feet per minute for dry machining. Cutter life is reported to be 3,000–5,000 lineal feet between grinds. The critical factor in machining is the rate of feed. If the feed is too slow, the cutter overheats and dulls faster; if it is too fast, a ragged edge may result. Another important factor in using carbide-tipped cutters with FRP is vibration. The possibility of vibration should be investigated if trouble is encountered in maintaining sharp tools. The milling machine should be firmly mounted on a solid footing. The machine should be overdesigned on the sturdy side rather than risk the danger of vibration during operation.

The cutter head could have 4–8 cutter blades, which blades should be designed with the edge of the cutter on the center line of the head having a negative rake of from 1° to 5° and a clearance or relief angle of 30°. The cutter should rotate into the stock to prevent delamination of the part. Deep slots can be milled if side milling cutters are used; otherwise, the FRP parts tend to bind the cutter.

Tapping and Threading

For tapping and threading, high-speed nitrated and chromium-plated taps are best. Ground taps are desirable. Three-fluted or two-fluted spiral gun taps are recommended. A negative rake of 5° on the front face of the land will help avoid binding and stripping of the thread when the tap is withdrawn. Taps should be 0.002–0.008 inches oversize since there is a tendency for the laminate to close up when the tap is withdrawn.

To prevent splitting, parts must be clamped when one is tapping parallel to continuous glass reinforcements. The hole edges should be chamfered and supported so that they will not be raised by the tap. The diameter of the chamfer should be the outside diameter of the tap with a 45–60° chamfer.

In production operations positive feeds are necessary for good tool life.

Mechanical Fasteners

Sometimes mechanical fasteners are required in an FRP assembly. Figure 2.15 shows a wide variety of mechanical fasteners that are available (although this list is not all-inclusive), a sketch of the fastener as received, and another showing it in place, with some explanations as to the suitability of the fastener for FRP applications. The most frequently used fasteners are self-threading screws and nuts and bolts. In automotive applications studs are the most popular fastener used.

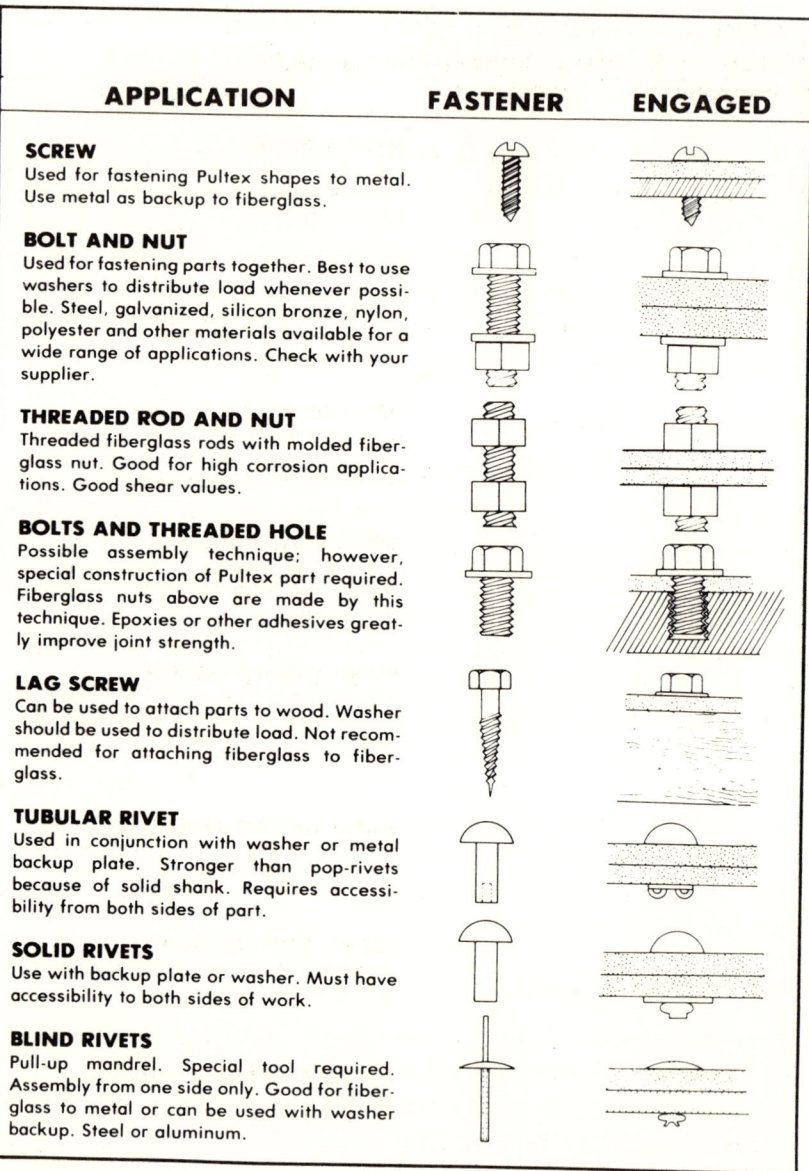

APPLICATION	FASTENER	ENGAGED

SCREW
Used for fastening Pultex shapes to metal. Use metal as backup to fiberglass.

BOLT AND NUT
Used for fastening parts together. Best to use washers to distribute load whenever possible. Steel, galvanized, silicon bronze, nylon, polyester and other materials available for a wide range of applications. Check with your supplier.

THREADED ROD AND NUT
Threaded fiberglass rods with molded fiberglass nut. Good for high corrosion applications. Good shear values.

BOLTS AND THREADED HOLE
Possible assembly technique; however, special construction of Pultex part required. Fiberglass nuts above are made by this technique. Epoxies or other adhesives greatly improve joint strength.

LAG SCREW
Can be used to attach parts to wood. Washer should be used to distribute load. Not recommended for attaching fiberglass to fiberglass.

TUBULAR RIVET
Used in conjunction with washer or metal backup plate. Stronger than pop-rivets because of solid shank. Requires accessibility from both sides of part.

SOLID RIVETS
Use with backup plate or washer. Must have accessibility to both sides of work.

BLIND RIVETS
Pull-up mandrel. Special tool required. Assembly from one side only. Good for fiberglass to metal or can be used with washer backup. Steel or aluminum.

Figure 2.15 Mechanical Fasteners (*Courtesy: Creative Pultrusions*)

FRP Part Design   41

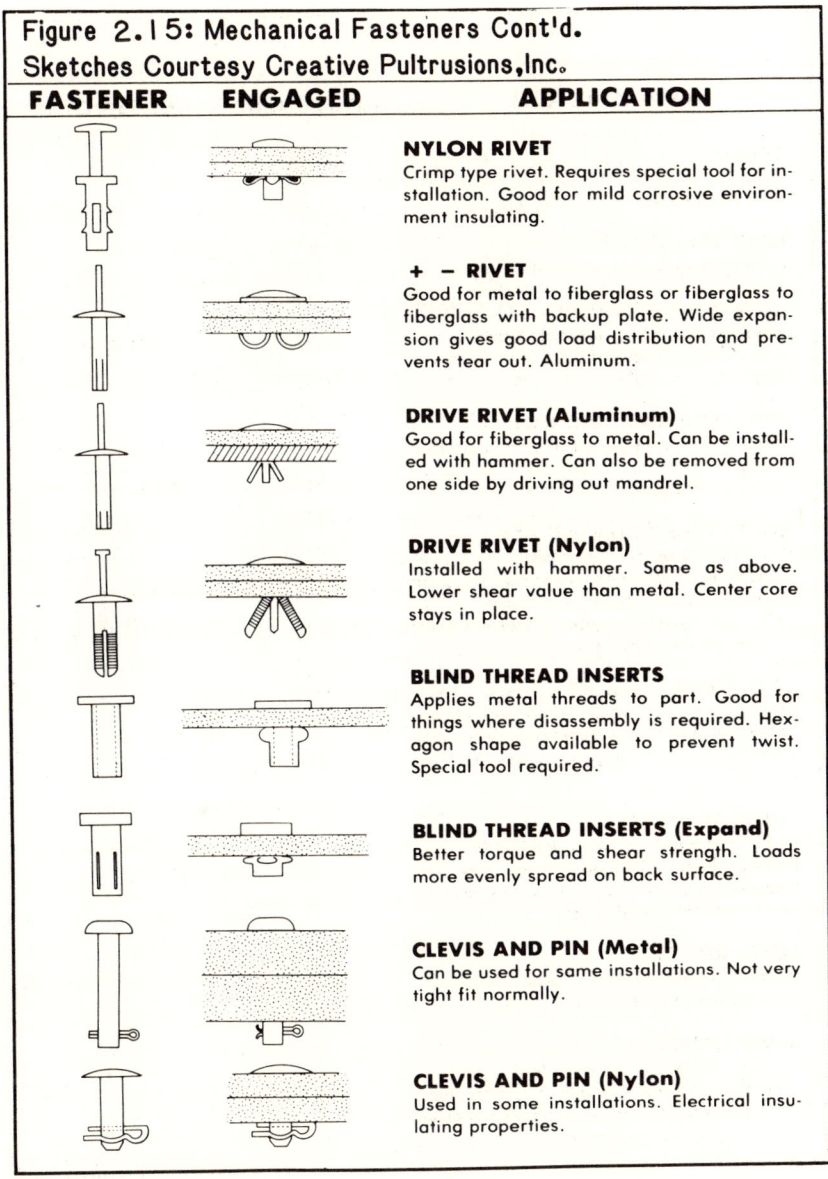

Figure 2.15: Mechanical Fasteners Cont'd. Sketches Courtesy Creative Pultrusions, Inc.

Adhesive Bonding

Adhesive bonding plays a major role in FRP assembly. It is recommended that an adhesive bond be connected with a mechanical fastener for optimum performance. That normally speeds up the assembly time, since postcuring the bond to develop maximum strength usually can be eliminated. The mechanical fastener will protect the adhesive bond until chemical action can develop full bond strength.

The following types of adhesives are used in FRP bonding: (1) epoxies; (2) urethanes; (3) acrylics; (4) polyesters.

A wide range of formulations is available, and it sometimes is a perplexing choice to select the one adhesive that is best suited for a particular job. Since any one adhesive will not fulfill all requirements, a variety of adhesives is usually available in an FRP assembly area. Epoxies are by far the most often used adhesive because of their easy availability, ease of application, and relatively low cost. The time to develop sufficient green strength so that the assembly can be removed from a bonding fixture is relatively long. For that reason the 2-part urethanes that develop good green strength very fast, especially in a heated assembly fixture, are preferred for automotive assembly work. Such adhesives are required in numerous automotive specifications. A relatively expensive dispenser is required to keep the two reactive components separated until the moment they are required in the assembly operation. Of course pressurized feed pots are needed, as is a static or dynamic mixer in the feed line, and a method of flushing the mixer and mix nozzle after an adhesive shot has been dispensed. One such dispenser is the Twinflow™ Automatic Resin System manufactured by Liquid Controls Corp. Other systems also are available. One of the more popular urethane systems for automotive applications is the Pliogrip™ FRP System manufactured by Goodyear.

Cyanoacrylate adhesives develop maximum strengths in minimum times at room temperature but are relatively expensive and much more difficult to handle.

Promoted unsaturated polyester adhesives are relatively easy to apply but require more time to develop sufficient green strength before they can be removed from a clamping fixture. They were the adhesive system specified for years in the assembly of the Corvette body panels but have been replaced by the

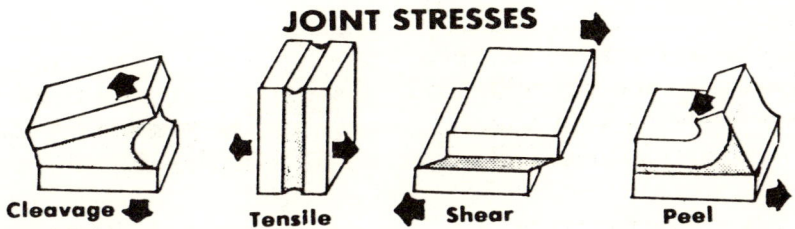

Figure 2.16 Examples of Joint Stresses (*Courtesy: Creative Pultrusions*)

urethanes. They remain valuable adhesives for nonautomotive use.

Surface preparation of the areas to be bonded is the most important step in adhesive bonding. Internal release agents are included in most FRP formulations, and most molds are sprayed with an external release agent before start-up of production. The residual release agent films that migrate to the FRP surface, any body oils from handling, and oil and grease contamination from nearby equipment must first be removed to obtain a good primary bond between two surfaces. A light sand blasting is the quickest method of surface preparation for most production operations. But when the two-part urethanes are used, a surface active agent can be applied that generally eliminates the necessity for sand blasting. A pigment or dye can be included in the surface active agent so that the operator can quickly determine, at a glance, that all surfaces have been coated. Usually there are die marks molded in on compression molded parts to outline the areas to be coated. The highlighted areas also are useful for quality control technicians during their routine inspections. A solvent, such as acetone, methyl ethyl ketone, or methylene chloride, also can be used to prepare the surface to be bonded.

The second most important factor in adhesive bonding is film thickness. That is not quite as important a factor for the urethane adhesives as it is for epoxy or polyester adhesives. In laboratory-prepared test specimens, adding glass beads of the diameter of the adhesive film thickness is a convenient method of controlling film thickness. In production bonding, controlling the film thickness must be built into the bonding fixtures.

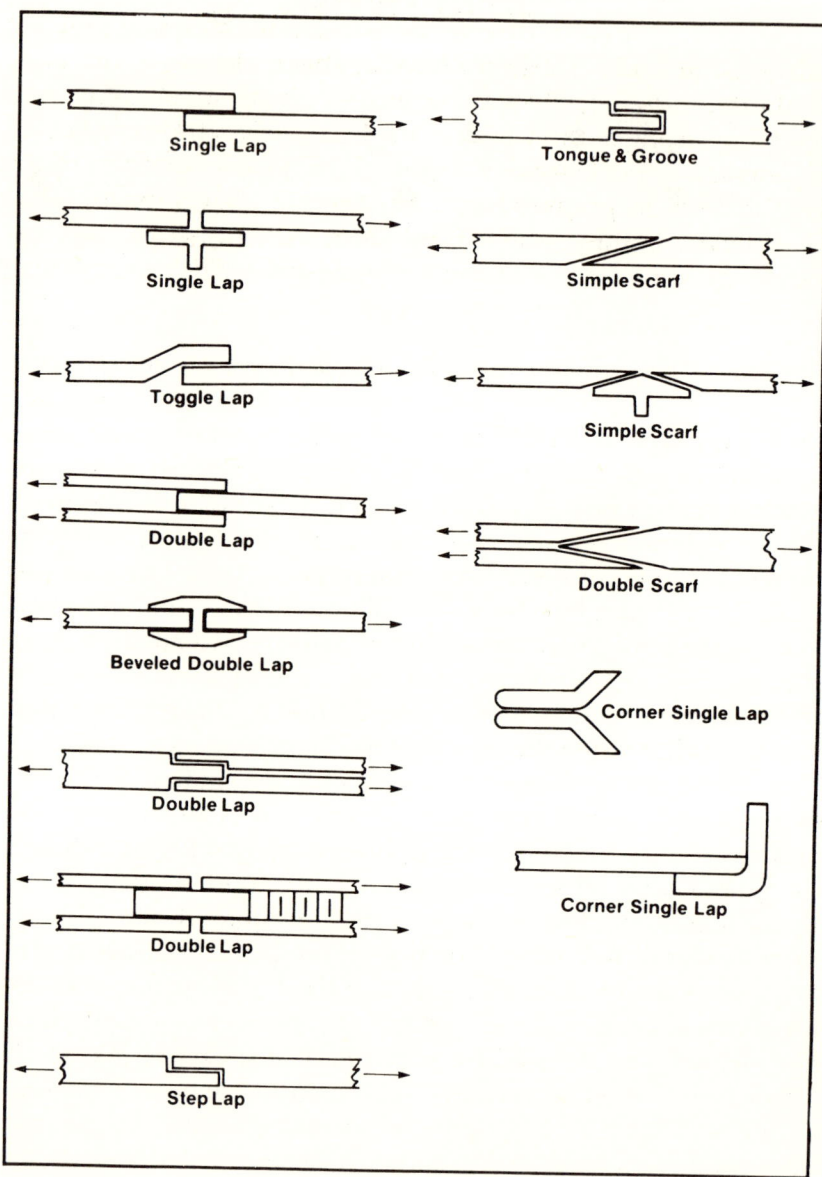

Figure 2.17 Several Bonded Shear Joint Designs (*Courtesy: Creative Pultrusions*)

There are four basic stresses on bonded joints when they are subjected to loads. These are tensile, shear, cleavage, and peel. Joints that are subjected to tension or shear stresses are considerably stronger and much more reliable than those subjected to cleavage or peel stresses. Figure 2.16 shows examples of the four stresses in graphic form. Figure 2.17 shows examples of 13 different bonded shear joint designs. The single lap and toggle lap joints are used most frequently in FRP assemblies.

CHAPTER 3

FRP Mold Design

General Information

It now is well understood that a good matched metal chrome-plated steel mold is required for optimum production conditions when one is molding bulk molding compound (BMC) or sheet molding compound (SMC) parts.

We will discuss the various details that must be covered in designing fiberglass reinforced plastic (FRP) matched metal steel molds. They have been presented in such a manner that they can be easily converted into a mold specification by the molder.

The actual design of the mold can be done either by the molder or by the moldmaker. Larger molders maintain their own engineering staffs to design molds, secondary fixtures, and to follow the mold during its progress through the moldmaker's shop. Smaller molders take advantage of the moldmaker's engineering staff for the development of working mold drawings. Either way it is necessary for the molder to develop a satisfactory mold specification sheet for the designer to follow.

The mold specification will automatically standardize such items as handling bolts, mounting means, heating channels, etc., so that most molds can be mounted in a press and made ready for production with a minimum of maintenance time.

When the moldmaker is expected to design the mold, the molder must furnish the following minimum information:

1. Part print
2. Master model
3. Size of press to be used with the mold, including mounting hole patterns, etc.
4. Expected molding pressure
5. Type of molding compound to be used
6. Molding material shrinkage
7. Ejector cylinder line pressure
8. Operating temperature ranges for both core and cavity
9. Type of steel to be used.

Figure 3.1 consists of a sketch of a typical small mold and includes the terminology used throughout this chapter.

Mold Steel

Preliminary mold drawings should have been completed by the responsible design group and approved by the molder before ordering steel.

The anticipated production quantities expected from the mold or the product end use or both should dictate the choice of mold steel, as shown in Table 3.1.

When steel forgings are used, a certification of the steel composition, vacuum degassing, and hardness should be supplied by the steel company to the moldmaker.

Stress relieving of both the core and the cavity are necessary after rough duplicating. The steel company should provide the time and temperature requirements for the proper stress relieving.

Master Models

Male master models of mahogany or laminated hardwood should be constructed to the outside surface of the part. Those master models should be finished with a protective coating to prevent moisture pickup or loss that might cause distortion.

FRP Mold Design   49

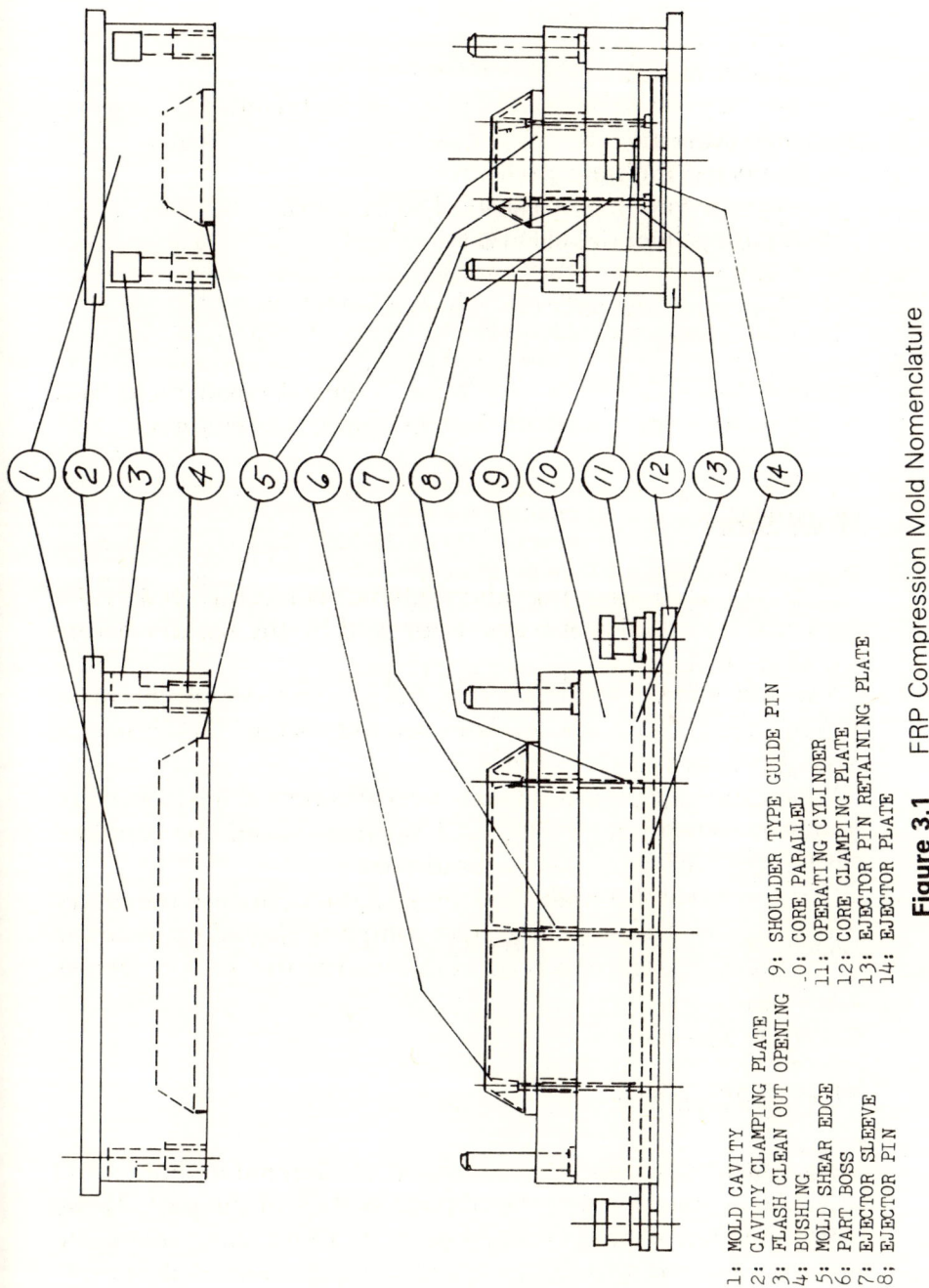

Figure 3.1 FRP Compression Mold Nomenclature

1: MOLD CAVITY
2: CAVITY CLAMPING PLATE
3: FLASH CLEAN OUT OPENING
4: BUSHING
5: MOLD SHEAR EDGE
6: PART BOSS
7: EJECTOR SLEEVE
8: EJECTOR PIN
9: SHOULDER TYPE GUIDE PIN
10: CORE PARALLEL
11: OPERATING CYLINDER
12: CORE CLAMPING PLATE
13: EJECTOR PIN RETAINING PLATE
14: EJECTOR PLATE

Table 3.1   Mold Steel Selector

	A. Production Planning Volumes	
	Type of Steel	
Planning Volumes	**Core**	**Cavity**
5,000–20,000 parts/yr	AISI*-1045 steel	AISI-1045 steel
20,000–30,000 parts/yr	AISI-4140 forged steel prehardened to Rockwell "C" of 28–32.	AISI-4140 forged steel prehardened to Rockwell "C" of 28–32.
Over 30,000 parts/year	AISI-4140 forged steel prehardened to Rockwell "C" of 28–32	P-20 forged steel prehardened to Rockwell "C" of 28–32.
100,000 parts or less for mold life	AISI-1045 steel	AISI-1045 steel
100,000–200,000 parts for mold life	AISI-4140 forged steel prehardened to Rockwell "C" of 28–32	AISI-4140 forged steel prehardened to Rockwell "C" of 28–32.
Over 200,000 parts during mold life	AISI-4140 forged steel prehardened to Rockwell "C" of 28–32.	P-20 forged steel prehardened to Rockwell "C" of 28–32.
	B. Product End Use	
Structural items where surface appearance is not critical, such as reinforcing panels, truck front ends, etc., where molded surface quality is of secondary importance.	AISI-1045 steel	AISI-1045 steel
High-quality surface appearance decorative items, such as grille opening panels, head lamp surroundings, quarter wheel opening covers, etc., where a high degree of polish is required on the outer part of cavity surface.	AISI-4140 forged steel	P-20 forged steel

*American Iron and Steel Institute

Frequently, the molder's customer furnishes the master model and the part print or mylar. Occasionally minor discrepancies exist between these two, and the customer should decide whether the part print of the model takes precedence. If the part print is to be followed, then the master model may have to be corrected, or vice versa. A master model should always be certified before construction of the duplicating models.

Duplicating Models

The master model is used to prepare a female glass–fabric-epoxy duplicating model. All model buildup by the moldmaker for parting lines, heel blocks, shear edges, mash-offs, etc., should be verified and approved by the molder before either cavity or core duplicating models are prepared.

Shear Edge Design and Specifications

The shear edges on an FRP mold differentiate it from other compression type molds. For SMC the design of the shear edges is particularly important. The purpose of the shear edges on an SMC mold is to provide a thin section at the extremities of the material flow that will cure quickly and then provide a dam to prevent further material escape, thereby establishing a back pressure on the curing material in the mold. As such, the shear edges no longer actually shear the glass reinforcement, as was the case peviously when oversized preforms had to be cut to size in the mold. Figure 3.2 consists of a sketch of both a horizontal and a vertical-type shear edge design. The following should be specified for SMC shear edges on molds:

1. Clearances should be from metal-to-metal fit to a maximum opening of 0.005 inches at the mold operating temperature of 300°F(149°C).
2. Shear edges on both the core and the cavity should be flame hardened to a Rockwell C of 55–60.

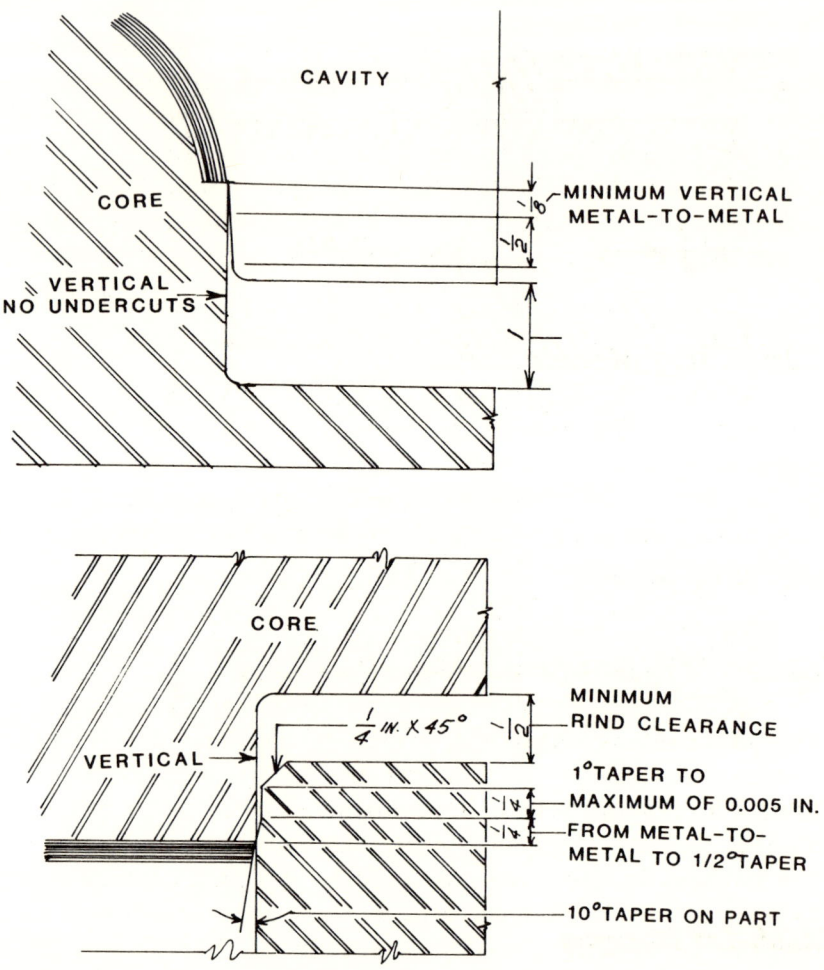

Figure 3.2   Shear Edge Designs

3. Final polishing should be in the direction of draw to prevent sticking or tearing of the flash. Shear edges should be polished well into the rind area.
4. Minimum rind clearances should be:
   a. horizontal shears—½ inch
   b. vertical shears—1 inch.

FRP Mold Design   53

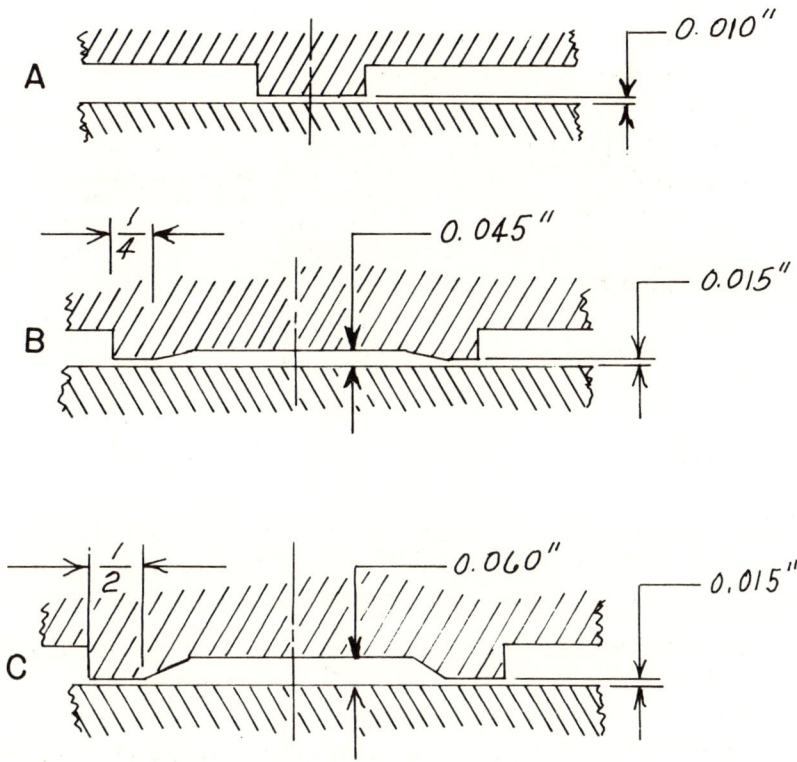

Figure 3.3   Mash-Off Area Designs

Mash-Off Designs

When it is necessary to include a mash-off area in a mold, the following designs are suggested:

1. Small Areas Less Than 1 Inch. For mash-off areas smaller than 1-inch diameter, square, or rectangular areas with sides of less than 1 inch, a uniform clearance of 0.010 inches is recommended, as shown in Figure 3.3A.
2. Medium-Sized Areas. For areas larger than 1 inch and smaller than 3 inches, an outside area with clearances of

0.010 inches with interior clearances of 0.045 inches is recommended, as shown in Figure 3.3B.
3. Large Areas. For areas larger than 3 inches, a pinch-off area of 0.010 inches with an internal clearance of 0.060 inches is recommended, as shown in Figure 3.3C.
4. Mash-off areas in a mold must be flame hardened to a Rockwell "C" of 50–55. Metal areas opposite the mash-off also must be flame hardened.
5. Mash-off areas should be highly polished to prevent material from sticking to the mold.

Heel Blocks

Molds should have heel blocks, hardened steel against bronze, suitable to withstand all lateral forces at 1,800 psi molding pressure. The following are recommended:

1. Heel blocks should be an integral part of the mold (not bolted on).
2. Heel blocks should be vertical.
3. Heel blocks should engage a minimum of 2 inches before the shear edges engage.
4. Heel blocks should have bronze wear plates on one side.
5. Wear plates should have a ⅛-inch minimum chamfer or radius lead in.
6. Contact areas opposite the bronze wear plates should be flame hardened to a Rockwell "C" of 50–55.
7. Grease fittings should be provided on wear plates with facilities to obtain grease from the outside of the mold if the heel blocks are not readily accessible with a brush.
8. Heel blocks must be balanced in area on opposite sides of the mold.
9. Wear plates should be bolted to the heel blocks with socket head cap screws and contain dowel pins to prevent shearing off the wear plates.

Internal Heating Conduits

1. All steam or oil lines, if at all possible, should be connected internally. Exterior looping should be avoided.
2. All steam lines should be ¾ inch (American standard taper pipe threads).
3. All inlets and outlets are to be permanently identified on the mold.
4. Heat line spacing in the core and cavity should be 4 inches center-to-center.
5. All pipe plugs should be steel hexagon socket type, and the threads should be wrapped with Teflon tape before installation.
6. A heating pipe pattern should be designed to provide a uniform temperature (within +/− 10° F) in both halves of the mold with the capability of heating to 320°F with 100 psig steam pressure.
7. All drilled lines should be cleared of all drill chips or other obstructions.

Leader Pins

1. Leader pins or guide pins should be provided on all molds, the diameter of which should be a minimum of 2% of the mold width plus length.
2. Guide pins should be of the shoulder-type construction to prevent their being pulled through the core mold half.
3. Guide pins should be installed on the core half of the mold and guide bushings on the cavity half of the mold.
4. All guide pins should be of identical height and at least ¼ inch longer than the highest point on the core, and their ends should be chamfered or tapered for lead-in.
5. The length of guide pin retention in the bushing should be at least 1.5 times pin diameter.
6. Guide pins should be set in from the outside of the mold by at least 1 pin diameter.

7. One pin should be offset to prevent misalignment of the core and cavity.
8. A clearance of 0.0005 inches per inch diameter should be provided.
9. Guide bushings should be the bronze-plated-on-steel type with figure-8 oil grooves.

Mold Stops

Mold stops are to be provided to limit the mold's vertical travel during closure. They should be the split type with a maximum thickness of 1 inch or ½ inch each mold half. Stops should be made of oil-hardened steel hardened to a Rockwell "C" of 55–60. All stops should be flat and contact each other simultaneously. They should be bolted to the mold with ¼-inch diameter steel socket head cap screws. Stops should be of sufficient area to withstand the full press tonnage should the press be closed without a charge in place.

Flash Blowout Openings

Openings must be provided for flash removal over the guide bushings at a point beyond the guide pin travel in the bushing. That opening should be at least 1.5 times the pin diameter.

Mold Operating Cylinders

Hydraulic cylinders normally are used to actuate ejector pins and slides in a mold. In specifying and installing hydraulic cylinders in FRP molds, the following should be observed:

1. All hydraulic cylinders should have rod ends cushioned at both ends, spring-loaded Teflon cups, and Teflon seals (such as Miller model H67B or equivalent).

FRP Mold Design   57

2. Hydraulic cylinders should not be mounted directly to the mold surface. A steel spacer should be used between the cylinder and the mold steel to minimize heat transfer.
3. Core pin cylinders and slide cylinders must be mounted in such a manner as to provide a continuous flash passage to the outside of the mold.
4. All piston rods should be keyed to prevent unintentional backout.
5. Port outlets on cylinders should be positioned to yield minimum maintenance hookup and servicing.
6. All ejector cylinders should be installed with linear alignment couplers (such as Hydro-Line) to reduce rod, seal, and bearing wear and to prevent binding or erratic movement caused by misalignment. A minimum of ⅛-inch clearance must be provided for the alignment coupler in the ejector plate.

Ejector Assembly

1. The core and cavity ejector pattern should be agreed upon between the molder and the moldmaker before the completion of the mold design and any steam line drilling.
2. All ejector plates must be solid steel. No welded construction is permissible.
3. Ejector retainer plates and ejector backup plates should be made of AISI-1020 hot rolled steel plate.
4. The top and bottom surfaces of all ejector plates should be Blanchard ground to within 0.005 inches parallel and flat before any drilling operations.
5. Positive return of the ejector plate must be provided for on all molds with return pins approximately 1 inch in diameter.
6. D-M-E type stop pins, or equivalent, must be provided on both sides of the ejector assembly.
7. Clearance between ejector pins and ejector sleeves should be a maximum of 0.0005 inches. In order to accomplish that fit, pins must be selectively picked from stock or plated.

8. Ejector assemblies are to be provided with a minimum of four ground and hardened guide pins and bushings.
9. Ejector plates are to be bolted together with socket head capscrews with the head on the clamp plate side of the assembly. Access to the capscrews must be provided for with clearance holes in the clamp plate.

Ejector Pins, Ejector Sleeves, and Core Pins

Ejector Pins

1. Use standard D-M-E, or equivalent, ejector pins.
2. The smallest diameter pin permissible is 5/16 inch. Pin length should not exceed 80 times pin diameter.
3. Ejector pins on contoured surfaces must be keyed to pervent rotation and to provide one-way assembly.
4. Pin lengths should not be altered more than ½ inch.
5. Head clearances on all moving ejector pins is shown in Figure 3.4.
6. Fit areas of ejector pins and ejector sleeves should be 5/8 inch.

Ejector Sleeves

1. Use standard D-M-E, or equivalent, ejector sleeves.
2. Ejector sleeves and ejector sleeve extensions must be solid. No welded construction is permissible.
3. All sleeves and sleeve extensions should be numbered and cross-referenced to their location in the mold with a like number.
4. The ejector sleeve internal edges and surface just below the edge where the fit surface ends must have a 60° included lead angle.

Core Pins

1. Use standard D-M-E, or equivalent, ejector pins.
2. The smallest diameter pin permissible is 5/16 inch. The pin length should not exceed 80 times pin diameter.

FRP Mold Design   59

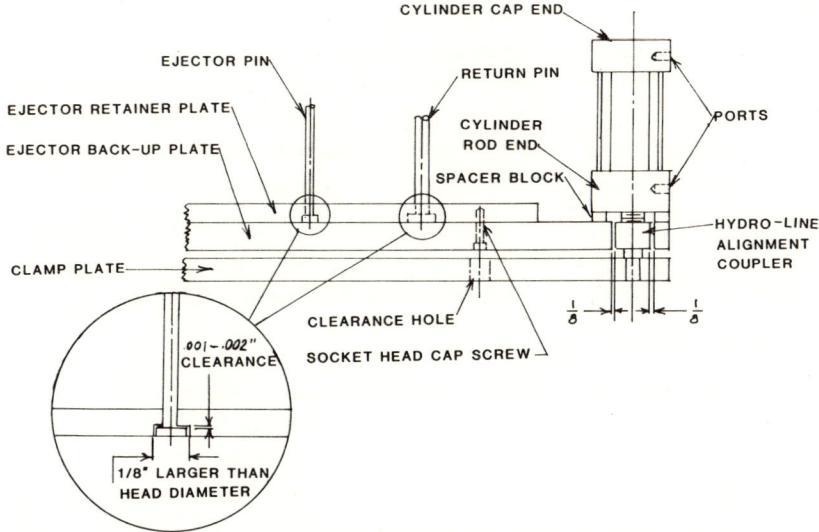

Figure 3.4   Ejector Plate Assembly Details

3. Pin lengths should not be altered more than ½ inch.
4. Head clearances on all moving core pins are shown in Figure 3.4.
5. Fit areas of ejector pins and ejector sleeves should be ⅝ inch.
6. Stationary core pins are to be held fast with a cover plate.
7. All core pins for holes in the line of draw should extend through to the rear clamp plate so they can be removed without affecting the balance of the mold. The head clearance counterbore should be ½-inch diameter minimum larger than the head diameter.
8. All pins are to be numbered and cross referenced to their location in the mold with a like number.
9. Pin diameter should be ¹⁄₁₆-inch minimum larger than pin tip. Core pin shoulder design should be according to Figure 3.5.

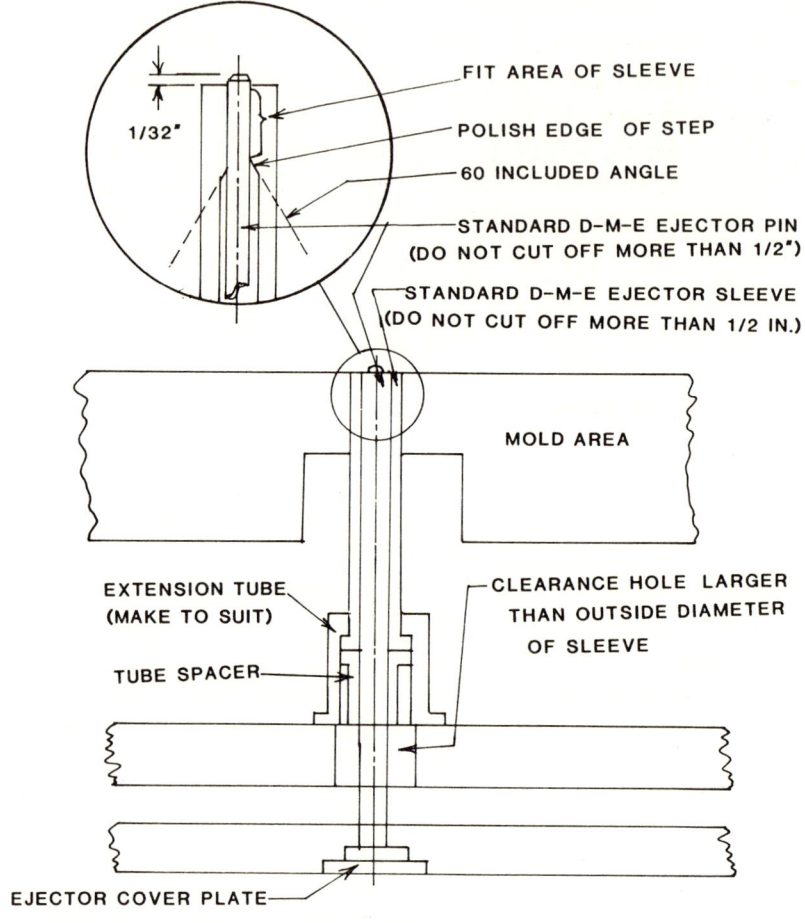

Figure 3.5 Stationary Core Pin Details

Handling Requirements

1. Eye bolt holes must be provided on all four sides of the cavity and core blocks and located with respect to the block center of gravity.
2. Thread diameters of the eye bolts on either mold block should be large enough to lift the entire mold.

3. All assemblies that weigh more than 200 pounds must have their own eye bolt holes for handling during mold maintenance.
4. Holes for tie straps for use during shipment and storage must be provided on two sides of the mold. The cavity and core should be blocked open a minimum of ½ inch with cold rolled steel plate before locating the tie strap holes.

Mold Surface Finish

Cavity

1. Appearance or show surfaces must be polished to 400 emery finish and buffed. No scratches, waviness, porosity, or other surface defects are permissible. On vertical surfaces the last polishing operation should be in the draw direction.
2. Nonappearance surfaces must be polished to 280 to 320 emery finish and buffed. Scratches, waviness, porosity, etc., are undesirable. The molder should indicate areas where those defects are unacceptable.

Core

The core must be polished to 280 to 320 emery and buffed.

General

1. All blind holes and bosses must be final polished in the mold draw direction. All electrical discharge machining (EDM) marks must be removed.
2. No undercuts are permissible.
3. Shear edges are to be polished and buffed in the mold draw direction to prevent flash from sticking to them.
4. Rind areas are to be polished and buffed a minimum two inches beyond the shear edge area.

Mold Tryout

Before a mold is accepted by the molder, it should be tried out. Larger moldmakers have presses and heat available for such tryouts in their shop or in a nearby area. When a mold is tried out for the first time in a moldmaker's shop, molder representatives should be present to witness the trial. The molder should provide the tryout or "break in" compound. A minimum of two sets of flash samples and of short shots for wall thickness checks should be made, one to be retained by the moldmaker and the other by the molder.

Mold Functional Check

Before the mold is shipped, the moldmaker should verify that:
1. All screws are secure.
2. The mold rests on all stops without pressure being applied.
3. The ejection system and all core pins operate satisfactorily.
4. Mold clamp plates are flat and parallel.
5. The shear is checked for backdraft.
6. All hydraulic cylinders operate without leaking.
7. All moving core pins retract beyond the part surface.

Chrome Plating

All BMC and SMC molds should be chrome plated in order to obtain the best possible surfaces on molded parts and to ensure maximum life to the mold surfaces.

It is recommended that both cavity and core be chrome plated, but only after the mold has been tried out and run long enough to verify that the surface finish is satisfactory, all units are functioning satisfactorily, and sample parts have been produced that meet the customer's specifications and end use requirements. Frequently a mold must be returned to the moldmaker to have minor rework in order to have a part meet all the print requirements.

The following apply to chrome plating:

1. Chrome to be 0.0006 to 0.0008 inches thick on flat areas.
2. Surfaces that are plated must be free of contaminants, rough spots, voids, arched areas, etc., which affect surface smoothness.
3. All chrome-plated areas must be buffed to a mirror finish after plating, with no streaks permitted. This should be completed, inspected, and approved before assembly of the mold.

BMC Specific Design Details

The mold design details mentioned previously in this chapter were written specifically to cover molds for SMC; however, most of the details supply equally well to BMC molds. Two areas that differ for BMC molds are the depth of mold travel after engaging the shear and loading wells.

1. Mold Shear Travel. Figure 3.2 shows two SMC shear edge designs that have telescoping lengths of ½–⅝ inches. The purpose of the telescope shear edge design is to allow the entrapped air to escape before pressure is put on the molding material inside the mold cavity. For BMC materials the total depth of this shear edge usually is ¼ inch. A ¹⁄₁₆ to ⅛-inch metal-to-metal contact generally is provided to obtain this seal followed by a ½° taper to the full ¼-inch depth. A ⅛-inch radius or a ¼ × 45° chamfer usually is included to protect the mold.
2. Loading Wells. As the glass content increases in BMC, so does the bulk factor. Normally no real problems are encountered with BMCs having glass contents of 10–20%. With glass contents of 25% and higher, the filler content must be reduced since the bulk factor increases rapidly. Bulk factor is the ratio of the volume occupied by the bulk material to that required for the molded part. Bulk factors

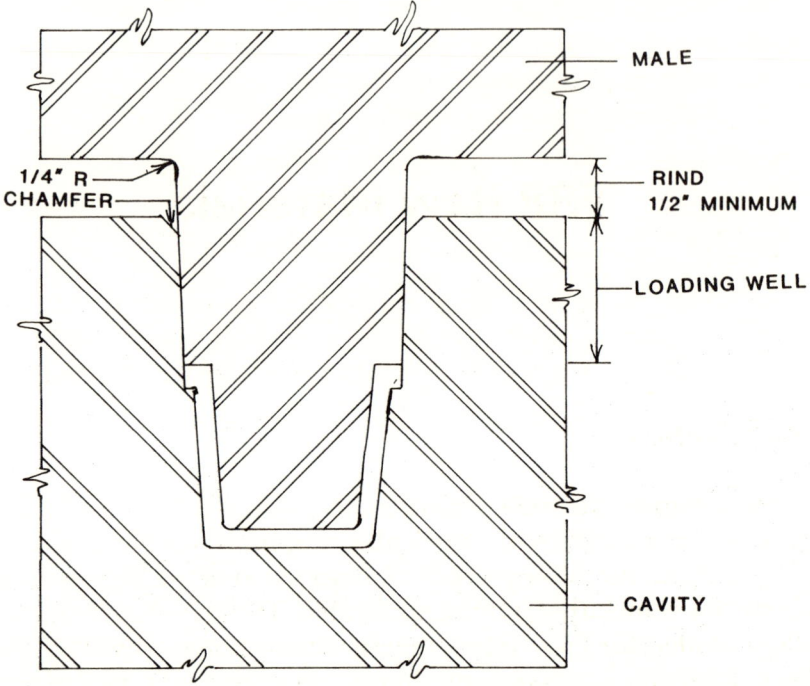

Figure 3.6 Loading Wells in FRP Molds for High Glass Content Compounds

usually are in the 2-to-4 range and can be as high as 7 to 8 to 1. In order to be able to contain the molding compound in the cavity of the open compression mold, a loading well must be included. Figure 3.6 shows the design of a mold for a high glass content cup that includes a loading well.

CHAPTER 4

FRP Raw Materials

Introduction

In this chapter, fiberglass reinforced plastics (FRP) raw materials are discussed with particular emphasis on their use in polyester molding compounds. A discussion of the background chemistry is included, where applicable. The chemical formulas are provided for inert fillers together with notes on where the base minerals normally are found, any relevant processing information, normal packaging, and the fillers contribution to the FRP product.

Schedules of commercially available products are included for polyester resins, thermoplastic additives, calcium carbonate, other inert fillers, glass roving, organic peroxides, and internal release agents.

Unsaturated Polyester Resins

The reaction between an acid and an alcohol produces an ester and water. By using difunctional alcohols or glycols and dibasic acids, esterification reactions can take place at each reactive site to form linear polymers. Generally, mixtures of unsaturated and saturated dibasic acids are used, but at least one of the acids must contain unsaturation. That esterification reaction does not affect the double bond or unsaturation in the dibasic acid. The chemistry of the reaction may be pictured as follows:

If G = Glycol
M = Maleic (or Fumaric) Unsaturated Acid
P = Phthalic (or other Saturated) Acid
S = Styrene (or other Monomer)

then the linear polyester may be represented as:
P-G-M-G-P-G-M-G-P

That polyester will not polymerize by itself; however, if another unsaturated material is included, such as styrene, that can react with the maleic unsaturation, then a cross-linked three-dimensional polymer can result. It may be represented as:

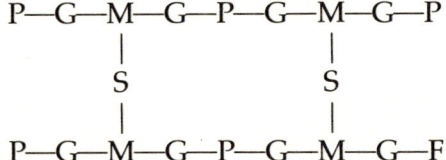

Different ingredients used in the cook bring different properties to the polyester resin. Table 4.1 contains a list of the more popular ingredients used to formulate polyester resins and the properties those ingredients impart to the polyester resin. Table 4.2 contains a listing and a description of the various types of polyester resins used by the FRP industry.

Table 4.1 Basic Building Blocks for Polyester Resins and the Properties They Bring to Formulations

Building Blocks	Ingredients	Characteristics
Unsaturated Anhydrides and Dibasic Acids	a. Maleic anhydride	a. Lowest cost, moderately high Heat Distortion Temperature (HDT)
	b. Fumaric acid	b. Highest reactivity highest HDT, more rigidity
Saturated Anhydrides and Dibasic Acids	a. Orthophthalic anhy.	a. Lowest cost, moderately high HDT, high flexural strength, high tensile strength
	b. Isophthalic acid	b. Higher tensile and flexural strengths, better chemical resistance, improved water resistance

Building Blocks	Ingredients	Characteristics
	c. Adipic acid, azaleic acid, sebasic acid	c. Flexibility (toughness, resilience).
	d. Chlorendic acid, tetrabromophthalic acid, hexachloro, octahydromethano naphthalene dicarboxylic acid	d. Flame retardancy
		e. Very high HDT
	e. Nadic methylanhydride	
Glycols	a. Propylene glycol	a. Lowest cost, good water resistance, flexibility, compatibility with styrene.
	b. Dipropylene glycol	b. Flexibility and toughness
	c. ethylene glycol	c. High heat resistance, good tensile strength, low cost
	d. Diethylene glycol	d. Good toughness, impact strength and flexibility
	e. Bisphenol-A	e. Corrosion resistance, high HDT, high physical properties
	f. Neopentyl glocol	f. Corrosion resistance, light color
	g. Trimethyl pentaneidiol	g. Corrosion resistance, lower reactivity
Monomers	a. Styrene	a. Lowest cost, high reactivity, good HDT, high flexural strength
	b. Vinyltoluene	b. Low volatility, more flexibility
	c. Methyl methacrylate	c. Light stability, good weatherability, fairly high HDT
	d. Diallyl phthalate	d. High heat resistance, long shelf life, low volatility
	e. Triallyl cyanurate	e. Very high HDT, high reactivity, high physical properties
	f. Monochlorostyrene	f. High reactivity, slight flame retardancy
	g. Alpha methyl styrene	g. Lower reactivity

Table 4.2 Description of Specific Polyester Resin Types

Types	Description
General Purpose	Standard rigid resin normally based on phthalic anhydride, maleic anhydride, and propylene glycol. Relatively good physical properties. Viscosity range: 500 to 4,000 cps
Low Profile	A mixture of a polyester resin and a thermoplastic additive. Can be supplied as a single component resin if the thermoplastic is compatible with the polyester/monomer solution
Light Stabilized	Excellent outdoor weather properties characterized by minimum discoloration to sunlight. Contains ultraviolet absorbers
Surfacing	Nonair inhibited type. Contains wax or other barriers to permit curing in the presence of air. Can be supplied with thixotropic additives for use on vertical surfaces. Coatings resist marring and blushing in water. Viscosity range: 500 to 3,500 centipoises
Lay-up	Formulated for ease of application and good glass wetting. Preaccelerated for hand lay-up or unpromoted for use with spray guns. Viscosity range: 250 to 3,000 cps
Chemical Resistant	Exhibits outstanding chemical and water resistance. Viscosity range: 500 to 1,000 cps or in solid form minus monomer
Heat Resistant	High heat distortion temperatures, higher hot strengths and rigidity. Viscosity range: 2,000 to 4,000 cps
Resilient	Greater toughness and impact strength. Some formulations have good hot strength. Viscosity range: 500 to 2,500 cps
Flame Resistant	Self-extinguishing, flame resistant, moderate heat resistance, and average strength properties. Viscosity range: 1,000 to 3,000 cps
Flexible	Low Barcol hardness, high flexibility and elongation. Viscosity range: 200 to 1,000 cps

Figure 4.1 is a schematic diagram of a batch process for unsaturated polyester resin production.

The type of polyester used in FRP systems is a highly reactive rigid resin containing much unsaturation. In fact, one of the more popular resins used for SMC automotive parts is an all-maleic-propylene glycol resin that by itself is very brittle. The thermoplastic additives, in addition to their primary function of providing superior surfaces, help partially plasticize the system to overcome some of that brittleness. Other formulations

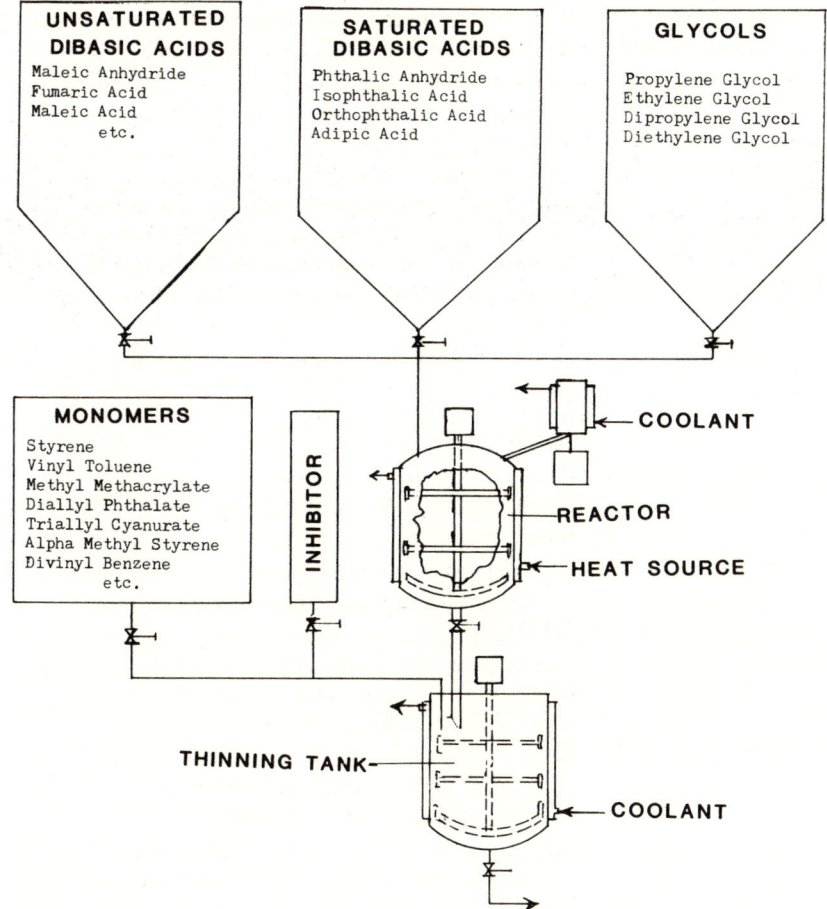

Figure 4.1 Schematic Diagram of Polyester Resin Production

include rubber compounds to toughen the system and overcome the brittleness. Other popular molding compound resins are very reactive rigid isophthalic resins.

Polyester resins are used in approximately 90% of all molding compounds. Both orthophthalic and isophthalic polyester resins are used in BMC. In older formulations the dibasic acid generally consisted of orthophthalic acid or anhydride with a mixture of propylene and dipropylene glycols. The saturated and unsaturated acid ratio was controlled to produce a medium reactivity resin. In low profile BMC a very highly reactive isophthalic resin generally is preferred, although the all-maleic resin used in SMC also is used. A small quantity of vinyl ester resins are used in molding compounds requiring good chemical resistance. In the early developmental stages of the reinforced plastics industry liquid phenolic resins were sparingly used because of their good electrical properties at very low cost. Phenolic resins seldom are used today in molding compounds. Triallyl cyanurate sometimes is used for high heat distortion requirements and diallyl phthalate is used on some electrical materials.

Table 4.3 contains a listing of commercially available polyester resins used for molding compounds. While the list is not all-inclusive, it does contain most of the important resins being currently used.

Polyester resins are shipped in 3,000-to-10,000-gallon bulk tankwagons or railcars and in 55-gallon lined steel drums. There generally is a 3-to-5-cent-per pound cost savings when one is purchasing in bulk. Most new molders purchase in drum quantities until their volume of use justifies installation of bulk handling facilities. Since the viscosity of the polyester resin solution for molding compounds generally is above 2,500 centipoises, it is not possible to pump or drain all the resin from a drum or tank. A minimum of 5 pounds, and more often 10 pounds, wall loss on the thicker resins remain after draining a 55-gallon drum of 500 pounds net weight, or a loss of 1–2%. That loss can often be completely eliminated when one is purchasing in bulk. Arrangements must be made to have an incoming tanktruck

Table 4.3 Commercially Available Polyester Resins for BMC and SMC

Resin Identification	Vendor	Monomer Type	Monomer %	Viscosity @ 25°C	SPI Cure Characteristics - Gel Time Min	SPI Cure Characteristics - Cure Time Min.	SPI Cure Characteristics - Peak Exo. °F	Thicken-able	Remarks
Altek™78–73	1	Sty	27	3,000	5–8	7–10½	390±10		High molecular weight DCPD resin
Altek™13–70	1	Sty	30	3,500	4–6	5½–8	440±10		Iso/Terephthalate resin
Aropol™2036	2	Sty	30	2,800	11	13	400	Yes	General-purpose BMC
Aropol™7030	2	Sty	38	1,200	5	6½	465	Yes	Low-shrink, pigmentable SMC
Aropol™7221	2	Sty	35	1,800	4	5½	420	Yes	Highly reactive, BMC
Aropol™7721	2	Sty	35	450	10	13	270		Low reactivity, flexible blending resin
Aropol™8110	2	Sty	27	3,500	5	7	390		Orthophthalic, general purpose, BMC
Aropol™50400	2	Sty	45	300	11	14	370	Yes	Low profile SMC, Phase II™
Aropol™50405	2	Sty	50	300	11	13	360	Yes	Low profile, thin skin SMC, Phase Alpha™
Hetron® 610	2	Sty	25	3,500	7	9	420		Fire-retardant concentrate, 30% bromine
Koplac™ 3702-5	3	Sty	39	500	28½	30½	450	Yes	High reactivity
Koplac™ 3102-5	3	Sty	34	550	19	21	410	Yes	Resilient resin
Koplac™ B-363-15	3	Sty	35	550	22½	24	415	Yes	Resilient BMC & SMC
Koplac™ 6002-5	3	Sty	40	550	4	5	460	Yes	BMC & SMC, Appliance parts

Table 4.3 Continued

Resin Identification	Vendor	Monomer Type	Monomer %	Viscosity @ 25°C	SPI Cure Characteristics Gel Time Min	SPI Cure Characteristics Cure Time Min.	SPI Cure Characteristics Peak Exo. °F	Thicken-able	Remarks
OCF-E-606	4	Sty	30	3,300	5½	8½	435	No	Tough BMC resin
OCF-E-933-2	4	Sty	33	2,250	9	10	450	Yes	Pigmentable
OCF-E-955-2	4	Sty	31	5,300	7	9	430	Yes	Tough, pigmentable, GP resin
OCF-E-966	4	Sty	35	2,400	5	7	430	No	Food grade resin
OCF-E-980	4	Sty	34	1,350	7	9	440	Yes	Structural grade resin
OCF-E-987	4	Sty	40	2,400	10½	12	460	Yes	Tough, structural grade resin
OCF-E-4297	4	Sty	50	1,200	5	7	520	Yes	Automotive grade, LP, one pack
OCF-CX-1631	4	Sty	50	900	5	6	500	Yes	Large body panel resin, one pack
OCF-RP-325	4	Sty	35	1,000	12	13	460	Yes	Use RP-325 with RP-500 and/or RP-775
OCF-RP-500	4	Sty	69	1,400	DNA	DNA	DNA	No	Low profile additive
OCF RP-775	4	Sty	38	1,300	DNA	DNA	DNA	Yes	LP additive, acid modified
Silmar-S-808HV*	5	Sty		3,500	7½	9	440	Yes	Modified BPA
Silmar-S-816P*	5	Sty		550	29	31	450	Yes	For Class A Surfaces
Silmar-S-817E*	5	Sty		2,750	6.6	8.4	470	Yes	Electrical applications
Silmar-S-832B	5	Sty		7,000	5	6½	450	Yes	Electrical appplications
Silmar S-846	5	None			DNA	DNA	DNA	No	Saturated polyester LP additive
Silmar S-866	5	Sty		550	29	31	450	SI	For use with reactive LP systems

Table 4.3 Continued

Resin Identification	Vendor	Monomer Type	Monomer %	Viscosity @ 25°C	SPI Cure Characteristics			Thicken-able	Remarks
					Gel Time Min	Cure Time Min.	Peak Exo. °F		
Stypol® 40-2057	6	DAP		23,000	11	14	330	No	Electrical grade BMC
Stypol® 40-2780	6	Sty		1,600	4	5	460	Yes	High reactivity, LS SMC
Stypol® 40-2782	6	Sty		1,150	6	7	415	Yes	High temperature, high reactivity
Stypol® 40-3919	6	Sty		450	7	8½	420	Yes	Expansion grade, tough SMC
Stypol® 40-8404	6	Sty		3,200	8	9½	400	No	Injection moldable BMC
Stypol® 40-3941	6	Sty		750	5	6½	450	Yes	Automotive grade
Stypol® 40-3949	6	Sty		2,300	7½	10	420	Yes	Abrasion resistant, high temperature resin
Stypol® 40-2643	6	Sty		1,200	8	10	400	Yes	Water resistant, high temp., excellent fatigue

Vendor Code: 1. Alpha Chemical Corporation
2. Ashland Chemical Company
3. Koppers Company
4. Owen-Corning Fiberglas Corporation
5. Similar Division, Vistron Corporation
6. Freeman Chemical Corporation
*Available in a range of viscosities and SPI Cure Characteristics

weighed at a local scale before and after unloading; the molder pays only for the net quantity of resin actually delivered. Otherwise, the normal 600-pound "heel" that remains in a 5,000-gallon tanktruck would be lost.

Brass, copper, and zinc should not be used in polyester piping systems. Copper can be either an inhibitor or a promoter, and zinc is a strong accelerator. Zinc die cast bungs in polyester drums have been known to be responsible for premature gelling of the resin during storage.

Other Resins

Vinyl Ester Resins

Vinyl ester chemistry has produced a resin that has chemical resistance and physical properties superior to polyesters, handling properties superior to both polyesters and epoxies, and that, at the same time, maintains a high degree of resiliency.

The resin in a reinforced plastics composite supplies the chemical resistance. If a chemical failure occurs, generally it is the resin that is attacked. In both polyesters and in vinyl esters, the failure occurs in the ester linkage, which is normally hydrolyzed. For many years the bisphenol fumarate polyesters have been accepted as having the best chemical resistance in the field. An examination of the molecular structure of both a bisphenol fumarate and a vinyl ester in Figure 4.2 will explain the chemical resistance improvement inherent in the latter resin. Note the recurring ester groups in the polyester. They appear in the main body of the chain and, once attacked, split the chain, leaving it susceptible to additional attack. Also note that the vinyl ester has only two ester linkages, both terminally located with the reactive vinyl sites (hence the name vinyl ester). Not only are there fewer ester groups to attack, but if they are attacked, they are terminally located and the main body of the molecule remains unaffected.

Most vinyl ester resins are used in hand lay-up work requiring superior chemical resistance. Small quantities of vinyl ester

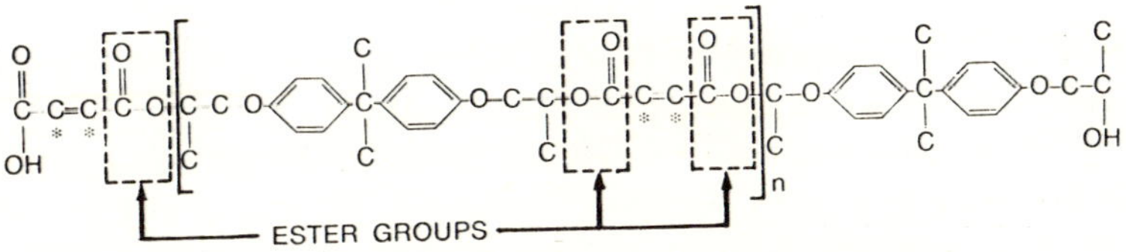

BISPHENOL-A/FUMARIC ACID POLYESTER

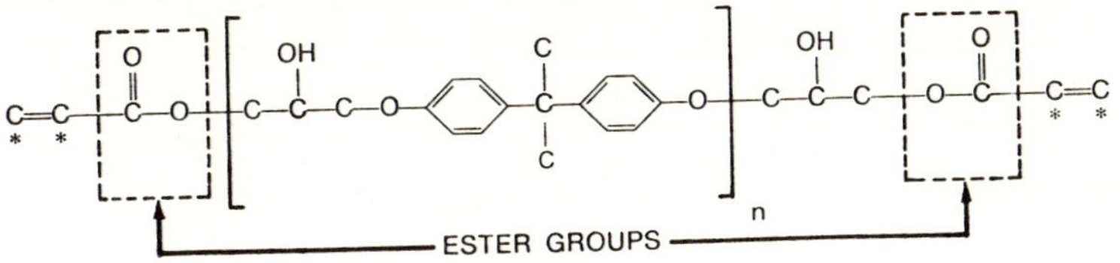

VINYL ESTER RESIN

*=REACTIVE GROUPS

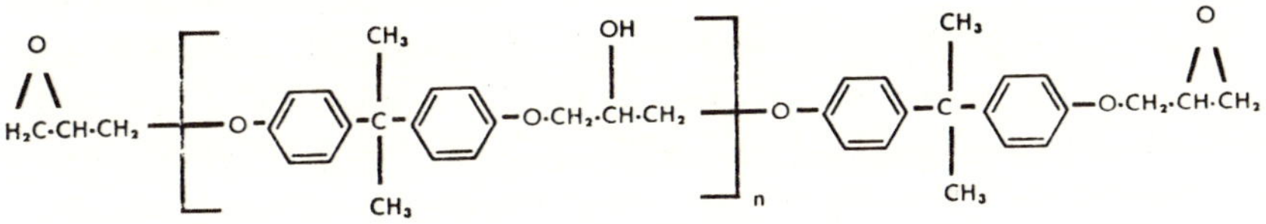

BISPHENOL-A/EPICHLOROHYDRIN EPOXY RESIN

Figure 4.2 Structural Formulations of Several Resins

resins are used in BMC and SMC. Increasing quantitites of vinyl ester resins are being used in other FRP parts requiring chemical resistance or toughness or both. Vinyl ester resins generally cost 25–35% more than unsaturated polyester resins.

Epoxy Resins

Epoxy resins possess unusually good electrical, chemical, and thermal properties. They exhibit very low shrinkage during cure and provide good adhesion to a variety of surfaces and materials. The type of epoxy used in most FRP applications is the bisphenol-A/epichlorohydrin type characterized by the general formula shown at the bottom of Figure 4.2. The value of n in the diagram is less than 1 for liquid resins and 2 or greater for solid resins.

Epoxy resins must be cured with hardeners (cross-linking agents) or catalysts to develop their desirable properties. The epoxy and hydroxyl groups are the reactive sites through which cross-linking occurs. Catalysts include amines, anhydrides, aldehyde condensation products, novalacs, and Lewis acid materials. For higher temperature cure and higher heat deflection temperatures, aromatic amines, such as methylenedianaline, are used.

Thermoplastic Additives

Most of the older lower-cost thermoplastic resins have found use as a low profile additive in an FRP system. Table 4.4 contains a list of some of the thermoplastic resins that have been successfully employed in an FRP system. The patent literature is more inclusive, but not all thermoplastics that can be used in an FRP system can be used economically.

Those thermoplastics that are soluble in monomeric styrene are charged to the formulation as a syrup. The usual solids content of such syrups also will be found in Table 4.4. Those thermoplastic syrups that do not contain carboxyl termination can be used as carrier materials for the Group II metal thickeners.

Those thermoplastic syrups that are compatible with the polyester resin (i.e., will not separate on standing) usually are mixed with the resin by the vendor and are sold as "one pack" or "one component" systems. An example of such a material is polyvinyl acetate. This eliminates the choice of thermoplastic family and the concentration of additive in the formulation by the molder. Most molders prefer to be able to vary both the type and concentration of thermoplastic additive in order to be able to control the compound shrinkage.

TABLE 4.4 An Outline of Thermoplastic Additives

Item	Thermoplastic Family	Percent Solids Content	Commercial Material Designation	Vendor	Carboxyl Termination	Non-Carboxyl Termination	Remarks
1	Acrylic	40	Paraplex P-681	2	x		Polymethylacrylate/acrylate copolymer
2	Acrylic	40	Paraplex P-701	2	x		Polymethylacrylate/acrylate copolymer
3	Polyethylene	100	Microthene FN-510	4		x	
4	Polystyrene	100	M9-C2	5		x	Medium-impact grade, pellet form
5	Polystyrene	35	LP-80	3		x	Medium-impact grade
6	Polycaprolactone	100	LPS-60	3	x		
7	Vinyl	35	LP-35	3	x		
8	Polyvinyl acetate	40	LP-40A	3	x		
9	Polyvinyl acetate	40	LP-90	3		x	
10	Cellulose acetate butyrate	100	EAB-551-0.08	1		x	
11	Cellulose acetate butyrate	100	EAB-551-0.2	1		x	
12	Cellulose acetate butyrate	100	EAB-551-0.1	1		x	

Vendor Code:
1. Eastman Chemical Co.
2. Owens-Corning Fiberglas Corp.
3. Union Carbide Corp.
4. U.S. Industrial Chemicals
5. Amoco Chemicals Corp.

Inorganic Fillers

The principle reasons for using inorganic fillers in an FRP formulation are[106] (1) reduction of polymerization shrinkage, and (2) regulation of compound plasticity. The fillers should be dry; that is, they should contain less than 0.5% free water content. They should be uniform and free from contamination. Some contaminants may cause localized reactions with peroxides or act as inhibitors. Nonuniform color or appearance in the molded part may also be caused by contaminants that are physically different from the filler.

Fillers are classified according to their particle sizes as coarse or fine. Coarse fillers are those having average particle sizes of 8 microns or larger. More specifically, they are the nonplatey or nonfibrous type having low surface areas, low oil absorptions, are easily wet by resins, can be very highly loaded, provide poor compound cohesiveness and localized resin-rich pockets, and tend to increase fiber agglomeration during molding. Those effects are predominately the result of large particle size, which causes filtering in the fibrous glass reinforcement and the large voids between coarse particles. Examples are calcium carbonate, anhydrous aluminum silicate, calcium metasilicate, silica, and coarse talcs.

Fine fillers are those having average particle sizes of 5 microns or less. Such fillers generally have high oil absorptions and high surface areas, induce high viscosities in formulations, provide a high order of cohesiveness, reduce the effect of heating upon the plasticity of the compound during molding, provide lubricity during mixing and molding, aid in reducing fiber agglomeration, and reduce localized shrinkages caused by a more uniform distribution of fillers. Examples of those types of fillers are kaolin clays, hydrous aluminum silicate, fine mica, fine talc, collodial silica, and precipitated calcium carbonate.

Table 4.5 contains an outline of various inorganic fillers that are used in FRP parts. Further discussion here will be limited only to those more widely used fillers together with their methods of production and their applications in FRP products. The

TABLE 4.5 An Outline of Inorganic Fillers for FRP

A. Calcium Carbonate (chalk, whiting, marble dust)
  1. Crystalline (domestic)(limestone, calcite)
     a. Wet ground
     b. Dry ground
  2. Amorphous chalk
     a. Imported (English, French, Belgian)
  3. Precipitated
     a. Uncoated
     b. Coated
B. Silicates
  1. Mica
  2. Aluminum silicate (Kaolin)
  3. Magnesium silicate
     a. Talc
  4. Calcium silicates
     a. Wollastonite
     b. Calcium metasilicate
  5. Aluminum magnesium silicates
  6. Sodium magnesium silicates
  7. Bentonite
C. Asbestos
  1. Chrysotile
  2. Anthopylitte
  3. Crocidolite
D. Silica Products
  1. Diatomaceous earth
  2. Quartz
  3. Ground silica
  4. Pyrogenic silica
E. Metallic Oxides
  1. Tabular alumina
  2. Alumina trihydrate
  3. Sapphire whiskers
  4. Titanium dioxide
  5. Iron oxide
F. Miscellaneous
  1. Barium sulfate
  2. Calcium sulfate
  3. Magnesium carbonate
  4. Powdered metals

properties contributed to molding compounds by various fillers are shown in Figure 4.3.

Calcium Carbonate

Chalk, limestone, and *whiting* are terms often used to describe calcium carbonate. Whiting can be either finely ground calcium carbonate prepared from chalk, limestone, or the product obtained by chemical precipitation from a solution or suspension containing lime. The latter product is most commonly referred to as a precipitated calcium carbonate.

Naturally occurring calcium carbonate, in a variety of crystalline forms and in all degrees of chemical purity, forms a substantial part of the earth's crust. It is difficult to determine

80   Handbook of Polyester Molding Compounds and Technology

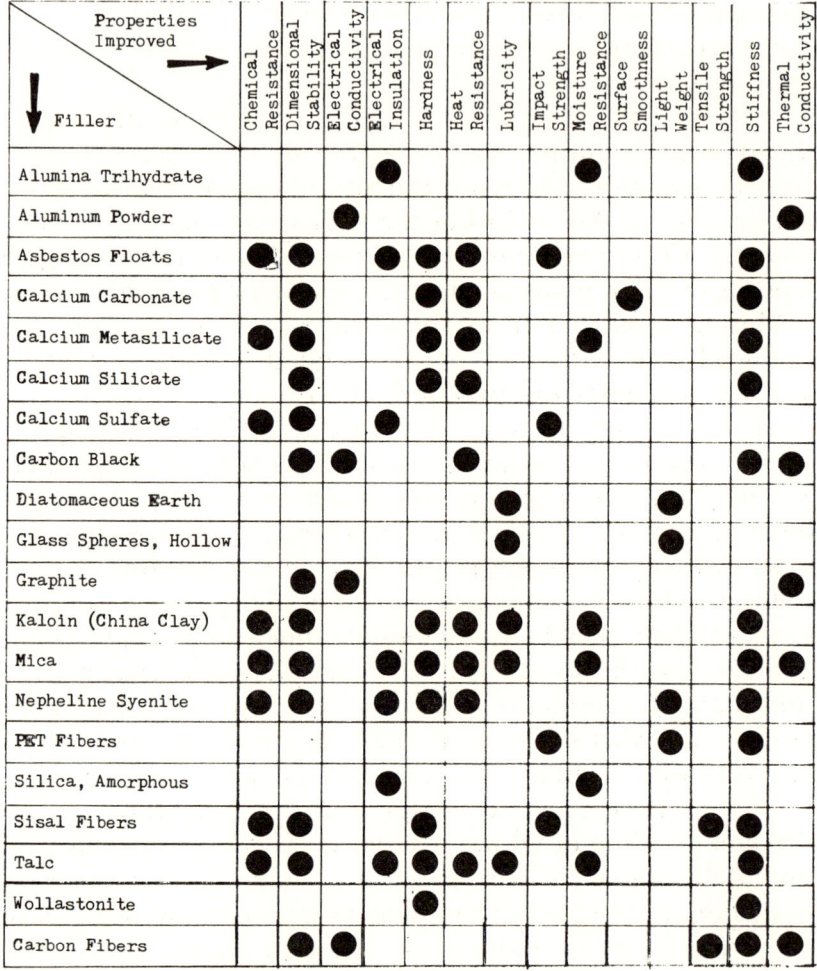

Figure 4.3   The Properties Imparted to Polyester Compounds by Fillers

exactly where a chalk grades into a marl or calcareous shale through increase of clay or silt and into a hard limestone through consolidation and crystallization because there are no rigid specifications for classifying them.

High-purity chalk occurs in Europe, particularly in England, France, Belgium, and Denmark. The United States chalk deposits, while similar in paleontological origin to those in Europe, are impure or off-color and cannot compare with the European chalks. Therefore, no whiting is now being produced from true domestic chalk.

Chalk is a natural calcium carbonate occurring as the remains of soft, friable, minute marine organisms, whereas limestone consists mainly of calcareous remains of organisms that inhabited oceans and lakes during the Cretaceous period. Geologically, the Cretaceous period received its name from the great deposits of chalk exposed on either side of the English Channel.

Whiting produced from chalk differs physically from that produced from limestone or precipitated whiting in that the particles are somewhat rounded rather than rhombohedral, as in calcite, and thus have greater surface area and consequently more absorptive power for comparable particle size. Both chalk and limestone are relatively soft, which makes the grinding of the material fairly simple. Ground whiting can be produced by either of two processes: dry grinding or wet grinding.

Dry grinding is less expensive than wet grinding, and it is done with ball, pebble, roller, and hammer mills and in impact pulverizers with closed circuit air separation. The latter method can produce a product of which 99.8% will pass through a 325-mesh sieve. Micronizer-type mills, in circulating particles at high velocity in a stream of air, reportedly can produce particles as small as 1 micron and averaging under 10 microns.

In wet grinding of whiting, continuous ball and pebble mills grind the limestone in the presence of water. It is then possible to fractionate according to particle size by passing the ground limestone slurry through continuous centrifuges. The various fractions are then filtered and dried in rotary, tray, or tunnel-type driers and then pulverized to break up the agglomerates that are formed during the drying step.

Precipitated whiting is a fine, white powder composed of microscopic crystals of calcium carbonate. It is odorless, tasteless, and nearly insoluble in water, and it is characterized by its small particle size and the absence of grit or other foreign matter. There are three processes used to manufacture precipitated calcium carbonate: carbonation, by-product, and calcium chloride. In each of the methods, limestone is used as the base material after it has been calcined to drive off the carbon dioxide. The lime is then slaked with water to form a milk of lime suspension, which is screened to remove impurities and coarse aggregates.

The by-product process is generally used by papermills to produce precipitated calcium carbonate for captive use. The product of that process requires considerable mechanical and chemical purification, so most manufacturers of precipitated calcium carbonate now use variations of the carbonation or calcium chloride processes. The carbonation process allows considerable latitude in control of temperature, concentration of reactants, rate of addition, and rate of mixing, but all of those factors may influence the particle-size distribution of the products. The calcium chloride process produces a calcium carbonate of high purity that is generally sold as U.S. Pharmacopeia (USP) grade.

Water extraction from all three processes is accomplished by rotary vacuum filters, which employ wash cycles to remove water-soluble salts. Another system concentrates the slurry in thickeners or centrifuges before filtering. The caked precipitated calcium carbonate contains 25–65% moisture, depending on the particle size, when it is removed from the filter. The moisture is removed by rotary, tunnel, or spray driers, and the dried material is disintegrated by passing it through a pulverizer. The powdered material is then conveyed to bins or direct to automatic packers.

Precipitated calcium carbonate is sometimes surface coated with fatty acids, resins, or wetting agents, which are applied either before or after the drying step. The treatment generally lowers the absorption characteristics, which allows more filler to be included in a mix at an equal viscosity but at a cost premium.

Precipitated calcium carbonate occurs in two crystalline forms, aragonite and calcite. The aragonite form is orthohombic in shape and has a specific gravity of 2.93, whereas the calcite has a specific gravity of 2.71 and is hexagonal and rhombohedral in shape. By control of temperature, rate of precipitation, and concentration of reactants, material may be prepared in which one or the other form predominates.

Calcium carbonate has a hardness of 3 on the Moh scale and is classified as a nonreinforcing filler. However, when used in concentrations greater than 10 parts per hundred parts resin, calcium carbonate improves the physical properties of polyester resins.

Calcium carbonate particles coated with stearic acid or resin are readily dispersible in polyester resins. It has been shown that the physical properties of the laminate are a function of particle size and particle-size distribution of the filler and are independent of surface treatment (Armstrong, 1964).

Most general purpose and automotive grade low profile SMC formulations contain up to 50% by weight of a fine water ground calcium carbonate whose particle size will average 2 microns. Increasing the filler concentration in a formulation will reduce cost, shrinkage, sink, flow, glass wet-out, specific volume, flexibility, flexural strength, tensile strength, elongation, and impact strength (PPG Industries Technical Bulletin).

Calcium carbonate should not be used in corrosion resistant and fire retardant applications. Table 4.6 is an outline of some commercially available calcium carbonate fillers.

Kaolin Clay

Kaolin (china clay) derives its name from the corruption of the Chinese kauling, the name of the hill near Janchau Fu, where the mineral was originally found. Kaolin is formed in nature by the decomposition of feldspar and occurs in many areas of the world. It occurs as primary deposits (found at the original point of formation) or as secondary deposits (transported by nature from the original point of formation and deposited in a new location). Large primary deposits of kaolin are found in England, but roughly 90% of the kaolin produced in the United States comes from secondary deposits in the Georgia-South Carolina area.

The ideal kaolin deposit is composed mainly of mineral kaolinite, which is a hydrated silica of alumina that may be

TABLE 4.6 Outline of Commercially Available Calcium Carbonates

Material Designation	Vendor	Oil Absorption	Average Particle Diameter Microns	Particle Range Microns	Remarks
Camel-Wite	1	15	3	0.2–12	
Gama-Plas™	2		7		Dry ground
Gama Co II™	2	15	1.5–1.9		Wet ground
Calwhite II™	2	13	1.9–2.2		Wet ground
Hi-Pflex 100	3	14	3	up to 15	Surface treated
Super-Pflex	3	26	0.5		Surface treated
Omya BLR/2	4	16	5	2–20	Dry ground
Atomite	5	15	2.5	up to 10	
Snowflake P.E.	5		5	1–12	
Hubercarb™ Q4	6	17	3.8		

Vendor Code: 1. Harry T. Campbell Sons Co.
2. Georgia Marble Co.
3. Pfizer Inc.
4. Omya, Inc.
5. Thompson-Weinman & Co.
6. J. M. Huber Corp.

represented by the formula $Al_2O_3 \cdot 2SiO_2 \cdot 2H_2O$. The alumina represents 39.5% of the total composition of kaolinite, with silica representing 46.5% and water 13.9%.

Most kaolin products contain small quantities of other minerals, such as feldspar, mica, and quartz, as well as trace quantities of iron, titanium, calcium, and sodium, etc. Kaolin particles are usually well formed, hexagonal flakes that occur in a wide distribution of sizes. Commercial products contain particles as large as 44 microns and as small as 0.1 micron expressed in terms of equivalent spherical diameter when measured by sedimentation in water. It is generally considered in the clay industry that kaolin particles below 2 microns are predominately single, flat, hexagonal plates, while clay particles coarser than 2 microns are strongly bound stacks or aggregates. Particles of less than 2 microns have a diameter-to-thickness ratio of about 8:1 but particles of larger than 2 microns are more nearly isometric in shape. Most authorities agree that stacks of aggregates or platelets are held together by hydrogen bonding.

Kaolin is mined from open pits after the overburden has been removed from the clay bed. The mineral kaolin beds vary from 5 to 40 feet in thickness, with up to 100 feet of overburden. As opposed to whiting, which must be mechanically ground to achieve small particle size, kaolin occurs in individual particles that are easily separated during the manufacturing process.

Air flotation and water classification are the two methods used to process kaolin clay. In the air-floatation process the crude kaolin is fed into a rotary kiln, where it is dried to less than 1% moisture content. From the drier it is pulverized and then fractioned with a whizzer-type cyclone. The oversized clay particles and the coarse impurities are discarded, and the kaolin is stored and prepared for loading and shipment. Since the properties of the final product are influenced by the properties of the crude clay, care must be taken and selective mining techniques are employed in order to ensure the particle size, whiteness, and brightness and impurity control of the final product.

The water-classification method of processing is a more detailed approach to the production of kaolin. The crude clay is suspended in water and degritted in the liquid state by wet cyclones and fine-mesh screens. The degritted product is subjected to centrifugal classification while still in water to produce products of different particle size. Each fraction is then processed separately through more screening, bleaching (to improve whiteness and brightness), thickening, filtering, rinsing, drying, pulverizing, and bagging operations. Wet process clays have much lower grit content, better color and brightness, and greater uniformity.

Calcined kaolin clay is produced by heat treating hydrated clay to remove the bound water. The process is accomplished with temperatures as high as 1,000°C. The calcination process alters the particle shape and particle-size distribution and increases the brightness and oil absorption of the product. The change in particle-size distribution is due to the agglomeration or fusion of the small particles of kaolin. Surface-treated kaolin coated with stearates and resins, which improve the wetting characteristics in organic systems, are produced.

Kaolin clays require more resin for wet-out and tend to adsorb or lose water as atmospheric changes occur, thus making thickening control more difficult and thereby limiting their use. Clays are sometimes used in FRP as a partial or complete replacement of the calcium carbonate filler to provide chemical resistance and to improve electrical properties of the laminate. Clays also impart a buff or tan color to the FRP product, generally making it more difficult to match colors, especially pastel colors.

Table 4.7 is an outline of several commercially available fillers other than calcium carbonate.

Talc

Talc is hydrous magnesium silicate represented by the formula $H_2O \cdot 3MgO \cdot 4SiO_2$ which corresponds to 4.8% water, 31.7% MgO, and 63.5% SiO_2. Talcs occur naturally as platelike and as fibrous needlelike particles. Unlike most other minerals, one portion of the talc particle is hydrophobic and resists wetting by water, and, therefore, it is easily wet by organics. The slippery or greasy feel of platey talcs results from each layer being

TABLE 4.7 Outline of Fillers Other Than Calcium Carbonate

Material Designation	Vendor	Type	Oil Absorption	Average Particle Size, microns	Specific Gravity	Remarks
McNamee Clay	1	Kaolin Clay		2		
Huber 35	2	Kaolin Clay	27	4	2.60	Water washed
ASP-400	3	Kaolin Clay	33	4.8	2.58	
Minex 4	4	Nepheline Syenite	22	7.5	2.61	
Snow White Filler	5	Calcium Sulfate	26.5	7	2.96	
GA-5	5	Calcium Sulfate	25.5	2	2.96	
C-331	6	Alumina Trihydrate		7.5	2.42	
SB-332	7	Alumina Trihydrate	27.5	11	2.42	
SB-432	7	Alumina Trihydrate	28	9	2.42	
G-431	7	Aluminia Trihydrate	28	9	2.42	

Vendor Code:
1. R. T. Vanderbilt Co., Inc.
2. J. M. Huber Co.
3. Englehard Minerals & Chemicals Co.
4. Indusmin Ltd.
5. United States Gypsum Co.
6. Aluminum Co. of America
7. Solem Industries Inc.

electrically neutral and adjacent layers being held together by relatively weak van der Walls forces. Therefore, when the material is rubbed together between the fingers, the weak bonds rupture and allow the layers to slide across one another.

New York and California are two of the leading talc-producing states. The New York product actually contains 25% of the mineral talc. Most of the product is tremolite–anthophyllite schist, somewhat altered to serpentine and talc. Other commercial talc deposits contain quartz, calcite, dolomite, magnesite, and limonite as well as many other impurities.

Talc is generally mined by underground methods, and the milling or particle-size reduction is usually accomplished by dry crushing, grinding, and air-classification methods. Talc is one of the most floatable minerals, and at least one process involves benefication in a water slurry. Most talcs are ground to 6-micron average particle size, but soft California talcs can produce products with 70% of the particles smaller than 2 microns in size.

Talcs are not generally used as the primary filler in an FRP formulation. They are mentioned here because it has been found that minor quantities of talc in combination with water-ground calcium carbonate can be used as a substitute for coated precipitated calcium carbonate in an FRP formulation. The talc apparently acts as a flow control agent. Talc also can be used to reduce the water absorption of molded parts.

Alumina Trihydrate

Alumina trihydrate, $Al_2O_3 \cdot 3H_2O$, is formed by precipitation from sodium aluminate liquors, and the average particle size can be closely controlled. The shape of the particles or crystals closely resembles particles of kaolin clay, and the particle size is similar to that of the finer kaolins, but the alumina trihydrates have a narrower particle-size distribution. Alpha alumina trihydrate produced commercially has a specific gravity of 2.40 and a refractive index of 1.57, and it contains 35% water of hydration that is evolved at 140°C and higher.

Connolly and Thornton (1965) have shown that general-purpose polyester resins filled with alumina trihydrate will provide equivalent flame resistance to a halogenated resin and antimony oxide system and at lower cost.

Alumina trihydrate is used as the principle filler in electrical compounds that require good arc resistance and good arc-tracking resistance.

Whenever high concentrations of alumina trihydrate are used in a compound, it is mandatory that the matched metal mold be chrome plated on both halves; otherwise the abrasive filler will wipe off quantities of metal from the mold surfaces into the compound as a contaminant.

Wollastonite

Wollastonite is a naturally occurring calcium metasilicate. It is the only commercially pure white mineral that is wholly acicular; typical length-to-diameter ratios range from 3:1 to 20:1. The average diameter of wollastonite is 3.5 microns.

The refined mineral is available in a range of grades and with several different surface modifications. NYCO offers grades modified with silanes, titanates, polymeric esters, etc. Wollastonite has been used for several years as a reinforcing filler in reaction injected molding (RIM), in reinforced reaction injection molding (RRIM), and in thermoplastic processing (Copeland and Rush, 1979). It was not until early 1981 that wollastonite found uses in the FRP industry.

Asbestos[1]

The best-known varieties of asbestos are chrysotile-$3MgO2PBSiO_2 \cdot 2H2$, amosite-$Fe \cdot MgSiO_3$, crocidolite-$NaFe(SiO_3)2FeSiO_3$, and tremolite-$CaO3MgO4SiO_3$. Of those several varieties, chrysotile was by far the most abundant, the

---

[1]*Note*: Asbestos has been included here, for historical purposes, since it was once an important ingredient in premix formulations. The use of asbestos is discouraged now, since it is considered to be a health hazard.

most widely processed, and the most widely used as a filler in plastics.

The bulk of asbestos fiber used in polyester molding compounds was of the floats classification. Those fine particles of asbestos fibers were obtained by an air-flotation process at the milling operation and varied in length from microscopic to 1/16 inch. That grade was designated 7TF.

The unique characteristics of asbestos floats as a filler in polyester molding compounds were:

1. its fibrous structure,
2. adjustable fiber length,
3. excellent retention of polyester resin during flow,
4. ease of mixing and molding,
5. high filler loadings possible.

The largest use of asbestos floats in polyester molding compounds was in automotive heater housings and automotive air conditioner ducts.

Asbestos floats in most formulations have been replaced by milled glass fibers, wollastonite, combinations of wollastonite and talc, and other inert fillers.

Miscellaneous Inert Fillers

Solid glass beads and hollow glass spheres have found application in both BMC and in SMC. Sundstrom, Collister, and Hayes[336] reported on the use of hollow glass bubbles to reduce the densities of BMC and SMC. Densities in the range of 1.2 to 1.4 g/cm^3 were reported, as compared with the customary values of from 1.7 to 2.0 g/cm^3. The bubbles have a bulk density of 0.24 g/cc with 80% of them being in the range of 20–95 microns diameter.

Nepheline syenite is a naturally occurring (Canada) anhydrous sodium potassium aluminum silicate that is of exceptional purity and exhibits low resin demand and extremely low tinting strength. Nepheline syenite is extremely easy to wet with polyester resins, and very high loadings are possible. Nepheline syenite contributes to overall electrical properties of parts.

Diatomaceous earth is a naturally occurring silica and is obtained from deposits of skeletons of organisms called diatoms. That form of silica has a hardness of 1-1.5 on the Moh scale. Diatomaceous earth is used to dry up molding compounds and to improve their flow.

Packaging

Calcium carbonate, china clays, wollastonite, etc., are usually packaged in 50-pound paper bags with 50 bags per skid or pallet. Some of the lighter fillers, such as asbestos floats, are packaged in bags of considerably less weight.

Larger molders purchase calcium carbonate in bulk rail cars. It is difficult to transfer and weigh bulk calcium carbonate and great care and judgment must be exercised in engineering bulk-handling systems. Most molders have relied on specialists to install their systems, and most such systems have had to go through lengthy debugging periods before they worked satisfactorily.

Fibrous Reinforcements

Roving

Continuous strand fiber glass roving is a collection of continuous filament glass strands. It is gathered without mechanical twist and is wound with even tension onto a cylindrical package on a universal winding machine. The strands forming the roving are in turn each composed of a group of continuous filament or individual glass fibers.

The process by which the glass filaments are formed involves molten glass flowing through tiny orifices at the bottom of a furnace. As the glass leaves the orifice, it is drawn or attenuated into filaments by means of a mechanical puller or winder at very high speeds, while a surface treatment or forming size is applied at a gathering station, which collects the filaments into a strand. The product at this point is known as a forming package.

The forming packages are dried to remove water and in some cases to cure the binder, and then they are transferred to a creel where the strands are drawn off simultaneously and wound on a roving package or "ball." The roving is wound on a paper tube, which is removed before shipment so that the package as supplied feeds from the center and weighs from 30 to 45 pounds.

Definition of Roving Terms

Filament Diameter Code.

A letter code generally is used in the roving product description to designate the diameter of the individual filaments in a strand. In the United States that diameter is expressed in hundred thousandths of an inch while in Europe the designation is in microns. Table 4.8 contains a breakdown of the letter codes and their diameters in both American and European units.

TABLE 4.8   Glass Filament Diameter Codes

Filament Diameter Code	Filament Diameter x 10^{-5} In.	Weight of 204 Filament Strand	Average Diameter Microns
B	10 to 14.9		
C	15 to 19.9		
D	20 to 24.9	450's	5
E	25 to 29.9	225's	7
F	30 to 34.9		8
G	35 to 39.9	150's	9
H	40 to 44.9		10
J	45 to 49.9		
K	50 to 54.9	75's	13
L	55 to 59.9		
M	60 to 64.9		
N	65 to 69.9		
P	70 to 74.9	37's	

Forming Size. The forming size contains the following ingredients: coupling agent, film former, lubricant, modifiers, and the carrier. The purpose of the forming size is to bond the individual filaments into a strand and to furnish protection to the strand during further processing.

The most important ingredient of the forming size is the coupling agent, which affords a chemical bond between the glass surface and the laminating resin functional groups. The name of the forming size is taken from the coupling agent employed, such as chrome (methacrylate chromic chloride), silane, or combinations of those, etc.

The most common film former used is polyvinyl acetate, but several others also are employed. Very little generally is mentioned regarding the nature of the film former used.

A lubricant (normally an organic vegetable oil) is employed to reduce frictional abrasion between the glass filaments and between the filaments and processing equipment.

In most sizes some modifiers are necessary, such as catalysts for reactive polymeric film formers, buffers to adjust the solution pH, etc.

Most sizes are aqueous solutions or suspensions, and distilled water accounts for 90–95% of most formulations. Some of the newer sizes contain solvent carriers, but care must be exercised in the solvent selection and application since the forming bushing at more than 1,000°C is within a few feet of the size applicator. The threat of fire and explosion has prevented more widespread use of solvent sizing systems.

Strands per Roving End. The quantity of strands in the roving describes the number of individual groups/ends of 204 (or 102 or 408) filaments each collected into the roving package. That was an important property early in the use of roving products. The designation now has no real meaning and is gradually being discarded.

Nominal Yield. The yield of a roving is the quantity of unit length contained in a unit weight. In the United States this unit is expressed as yards per pound and in Europe as meters per kilogram. Since the yield of roving products may vary considerably, it is advisable to verify the yield of each shipment and sometimes to make compensation by changing the cutter speed in order to maintain good glass content control.

Handling Type. The terms *hard* and *soft* relevant to rovings have come into general use throughout the FRP industry in the United States. They refer primarily to the tendency of the individual roving strands either to retain their integrity (hardness) or to open readily when the roving is bent around a pin giving a high degree of filamentation (softness). The hardness or softness of roving is achieved by changes in the sizing formulation and by carefully controlling winding and drying conditions.

Specification Properties

In addition to the yield, which was mentioned earlier, moisture content, and ignition loss are generally specified in the vendor's literature.

Moisture Content. A sample of roving (10 yards) is wrapped into a convenient hank and tied together. After weighing it is placed in an air-circulating oven for 1 hour at 105°C, cooled in a dessicator, and reweighed. The percent moisture loss is calculated. Moisture contents of continuous rovings range from 0.005 to 0.3%.

Ignition Loss. The previously dried sample used above is placed in a muffle furnace for 1 hour at 615+/−25°C, cooled in a dessicator, and reweighed. The percent ignition loss is calculated. Ignition losses on continuous rovings range from about 0.5 to nearly 3.0%.

Other Properties. In some vendor literature other roving properties, such as "fuzz," "ribbonization," and "stiffness," are mentioned. All those properties are important for the preform process, but only fuzz has a direct bearing on SMC roving. Fuzz balls may collect during chopping of the roving and drop into the SMC product. That can cause unnecessary rejects in molded parts, since fuzz balls generally will not wet out. The number expressed for fuzz is the weight in grams of filaments collected when the roving is pulled around sharp corners in a fuzz-testing machine for a given period. The lower the quantity of fuzz or filaments collected, the better is that roving for SMC use.

Handling Roving

To prevent contamination, roving should be received, stored, and moved within the plant in the vendor's shipping carton. Roving should be stored in a dry warehouse to prevent moisture pickup. It is recommended that roving be purchased in Creel-Paks™ or Stak-Paks™ (PPG) and be used directly from those pallets. In most such shipping containers the three balls of roving in a vertical cell are tied together and usually need no attention after start-up until the three balls have been completely used.

Stainless steel or nylon tubes can be used to lead the roving from its shipping package to the cutter location. Stainless steel is best since it can be grounded to dissipate most of the static charge on the roving package or generated during the pulling operation. Aluminum and mild steel tubes are unsatisfactory. Steel will rust, and the abrasive roving will cut through the aluminum in quick order and contaminate the SMC product with black aluminum oxide streaks. To start the roving through a tube, one must feed the end into the tube, and a short blast of air from an air nozzle will carry the roving completely through the tube.

If individual cartons of roving are used, they should be placed on a bookcase-type shelf after first removing the top carton flaps. A guide eye should be placed over the center of each package. It will be necessary to have the shelves approximately 18–20 inches apart to allow space for the roving to swirl as it unwinds from the package.

Table 4.9 contains an outline of the various rovings that are available for SMC use and the principle characteristics of each.

TABLE 4.9 Outline of Rovings for SMC

Vendor	Vendor Designation	Yield Yds/lb	Ignition Loss, %	Strand Type	Binder Compatibility	Binder Type
OCF	433	113		PE, VE	Silane	
OCF	433	227			PE, VE	Silane
OCF	951	113		PE	Silane	
OCF	956	113			PE, VE	Silane
PPG	515	119	2.0	ECK	PE	Silane
PPG	515	239	2.0	ECK	PE	Silane
PPG	516	119	1.05	ECK	PE	Silane
PPG	516	239	2.05	ECK	PE	Silane
PPG	521	112	2.1	ECK	PE	Silane
CT	235–04	110	1.5		PE	Silane
CT	255–B4	210	1.5		PE	Silane

Reinforcing Mats

There are two general types of fiberglass reinforcing mats: chopped strand mat and continuous strand mat. Both offer essentially the same degree of nondirectional reinforcement but have different molding and handling characteristics. Chopped strand mat is available in many thicknesses, described in terms of weight per unit area (weight in ounces per square foot or grams per square meter), ranging from ¾ to 4 ounces per square foot in the United States. Stiffness, wash resistance, and color of reinforcing mats depend upon the binder used to hold the strands together. There are two general binder types: soluble and insoluble (in styrene monomer).

In SMC, chopped strand mats are used having soluble binders in weights of from 1½ to 2 ounces per square foot. Table 4.10 contains an outline of several such reinforcing mats that are available.

Continuous strand mats generally are not used in SMC since they will not flow. SMCs made with continuous strand mat have been used for special applications as reinforcements for knit lines and as the center ply in a balanced ply-up for large-area flat parts. In some moldings pieces of continuous strand

TABLE 4.10 An Outline of Fiberglass Reinforcing Mats

Mat Designation	Vendor	Type Mat	Mat Densities Available In Oz./Sq. Ft.	Widths Available in Inches	Remarks
M127	1	Chopped strand	1½, 2, 3	4, 6, 8, 10, 12, 38, 50, 60, 76, 38, 50, 60, 76	
Hybon® AKM	2	Chopped strand	1 to 3 in ¼ oz. increments.	12¼ to 120 in ¼ inch increments	
M8605	3	Continuous strand	1 to 3 in ¼ oz. increments	14 through 72 in ½ inch increments	
M8680 M8681	3	Continous strand	¾ to 3 oz. in ¼ oz. increments	14 through 72 in ½ inch increments	Standard compression molding mat. Faster wet through for laminates over ¼ inch thick where surface smoothness is not critical.

Vendor Code: 1. CertainTeed
2. PPG Industries
3. Owens-Corning Fiberglas Corp.

mat are used to reinforce local areas where severe flow lines cannot be avoided. One such part is large concrete forming domes where the four vertical edges and the four corners need additional reinforcement.

Chopped Strands. Chopped strands in ¼- and ½-inch lengths are used as the reinforcement in premix and BMC. Generally most physical properties are directly proportional to the weight % of glass content in a compound, provided that the glass has not been damaged or destroyed by overmixing. Table 4.11 is an outline of fiberglass chopped strands.

Glass lengths of ⅛-inch are used in many thermoplastic compounds and in some thermoset compounds intended for injection molding. Using lengths in excess of ½ inch merely makes it more difficult to wet out the glass strands without contributing much in the way of additional properties. Those longer lengths generally are broken down in the mixing operation. Figure 4.4 shows the relationship of fiber length and concentration to the Izod impact strength of SMC.

Figure 4.5A shows the effect of both "insoluble" and "soluble" glass types and glass content on SMC impact strength, and Figure 4.5B shows the effect of glass content on tensile strength.

Organic Peroxides

Organic peroxides are useful as initiators or cross-linking agents because of the thermally unstable O-O bond that decomposes to form free radicals. Organic peroxides may be viewed as derivatives of hydrogen peroxide in which one or both hydrogens are replaced by organic radicals. Table 4.12 lists the structures of several commercially available organic peroxides (see p. 103).

The rate at which a peroxide decomposes into free radicals is dependent on temperature. As the temperature increases, the peroxide decomposition rate increases. Since polymerization reactions take place at various temperatures, organic peroxides have been developed with different decomposition rates,

TABLE 4.11 An Outline of Fiberglass Chopped Strands

Vendor Designation	Vendor	Available Lengths Inches	Binder Content In %	Binder Type	Remarks
Type 450	1	1/8, 1/4, 1/2	1.30	Silane	
Type 1156	1	1/8, 1/4, 1/2	1.15	Silane	
Type 3303	1	1/8, 1/4, 1/2	2.0	Silane	
Type 3029	1	1/8, 3/16, 1/4, 1/2		Silane	
832BB	2	1/4, 1/2	2.10	Silane	Insoluble binder designed for BMC and premix compounds
832BE	2	1/8, 1/4	2.10	Silane	
847GE	2	1/8, 1/4	2.20	Silane	Designed for electrical molding compounds
847HE	2	1/8, 1/4	2.20	Silane	
405AA	2	1/8, 1/4	1.91	Silane	Used in general-purpose and low-profile compounds
405AB	2	1/4	1.85	Silane	
405AC	2	1/2	1.85	Silane	
927	3	1/8, 1/4, 1/2	0.75		Microwave cookware applications

Vendor Code: 1. PPG Industries
2. Owens-Corning Fiberglas Corp.
3. CertainTeed

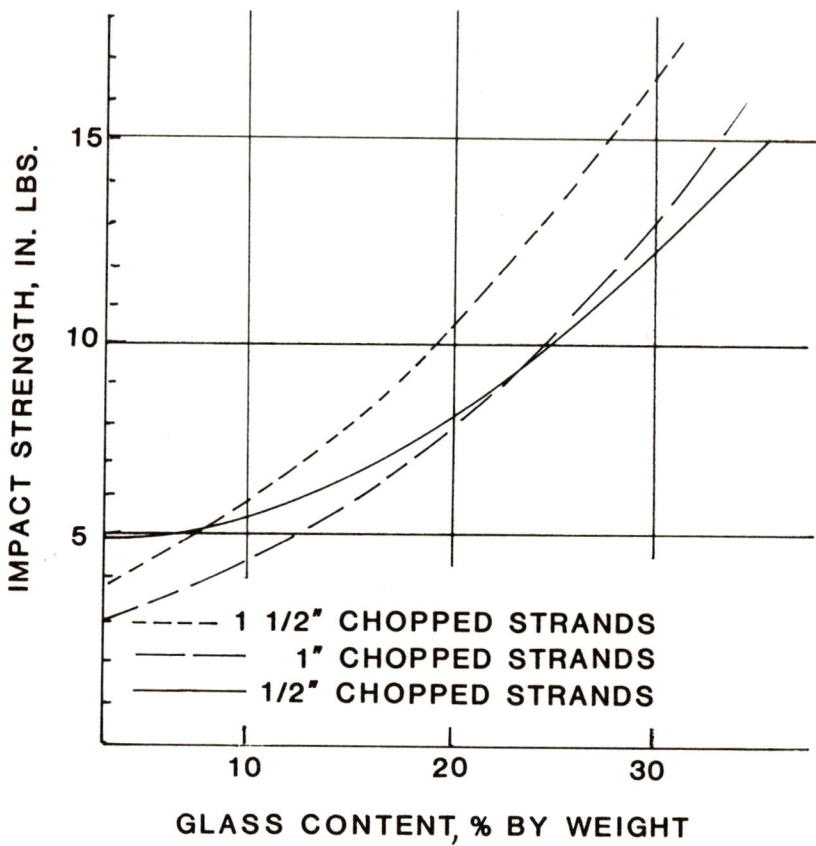

Figure 4.4 The Effect of Glass Length and Concentration on the Impact Strength of SMC

and they can be conveniently expressed by half-lives (the time required to decompose 50% of the peroxide in a diluent at a given temperature).

Table 4.13 (see p. 104) consists of an outline of the more popular organic peroxides used in BMC and SMC together with their chemical names, half-life data, and useful molding temperature range. Probably the most used organic peroxide in SMC is tertiary butyl perbenzoate (TBPB).

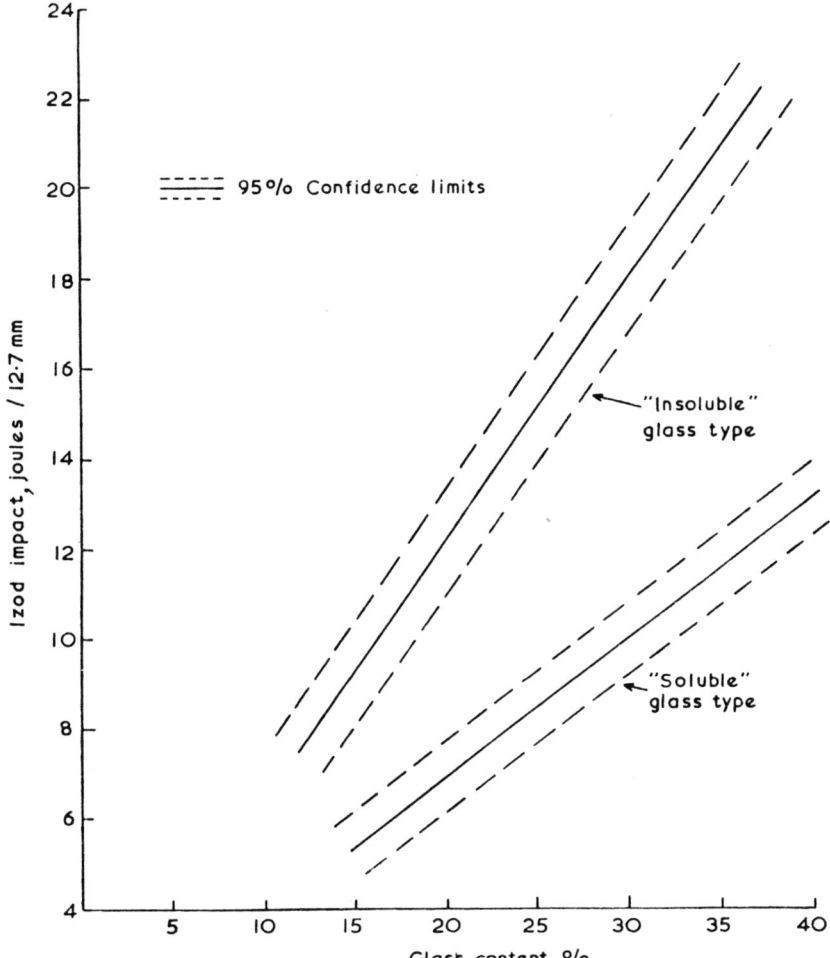

Figure 4.5A Impact Strength of BMC Versus Glass Content (*Source*: Burns, *Polyester Molding Compounds*, Marcel Dekker, 1982)

Figure 4.6 consists of the tracings of a series of Society of the Plastics Industry (SPI) Cure Characteristics tests plotted on top of each other. The resin is a popular isophthalic polyester containing several concentrations of various organic peroxides and

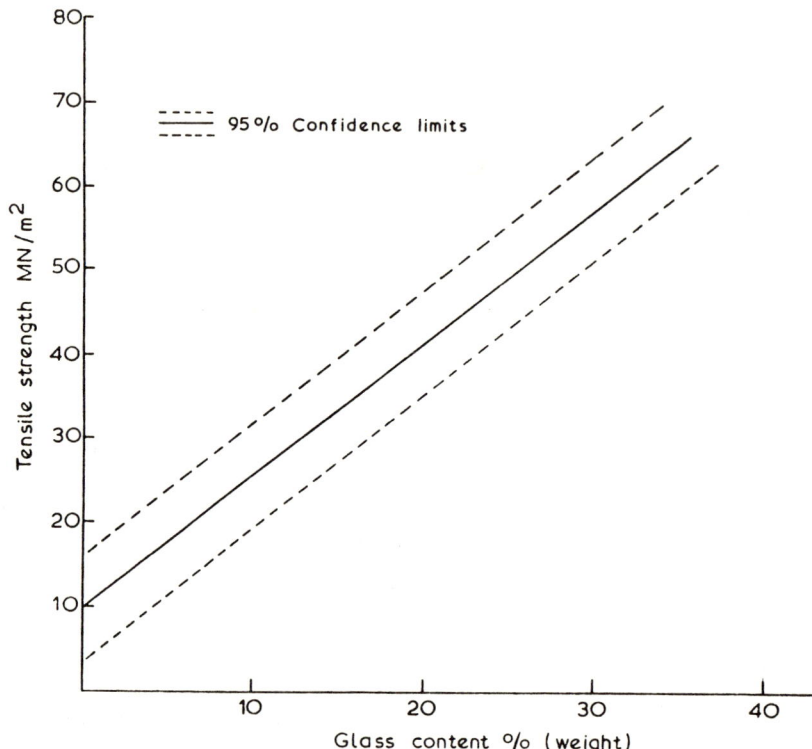

Figure 4.5B Tensile Strength Versus Glass Content (*Source*: Burns, *Polyester Molding Compounds*, Marcel Dekker, 1982)

combinations of peroxides. The test was run at an elevated temperature to more closely approach molding conditions. That is a relatively simple laboratory procedure for comparing catalyst systems. By examining the curves, it is easy to see that the higher temperature catalysts give a more sharply pronounced peak exotherm than do the lower-temperature catalysts, such as benzoyl peroxide. By measuring the time required to reach the peak exotherm from any point on the time axis, one can get a reliable comparison of the speed of the various systems.

Figure 4.7 contains a plot of the catalyst concentration (TBPB) versus the spiral flow length in inches of a general purpose BMC. As the catalyst concentration is increased, the flow length

TABLE 4.12 Structures of Commercial Organic Peroxides

Hydrogen Peroxide	H-OO-H
Hydroperoxides	R-OO-H
Dialkyl Peroxides	R-OO-R
Diacyl Peroxides	R-C(O)-OO-E(o)-R
Peroxyesters	R-C(O)-OO-R
Peroxy Acids	R-C(O)-OO-H
Peroxy Ketals	R_2-G-OO-R_2
Peroxy Dicarbonates	R-OC(O)-OO-C(O)O-R

of the compound decreases. The diagram explains why too much catalyst concentration in a compound can lead to increased rejects due to insufficient flow.

Internal Release Agents

Lubricants or internal release agents must meet the following requirements: they must be compatible during processing, not adversely affect physical properties of the end product, not contribute undesirable color or color drift, and be easily added to the resin mix. The selection of a suitable lubricant becomes a compromise with the total resin system.

Compatibility or solubility of lubricants in a resin system is dependent upon the melting point of the lubricant, the molecular structure (polarity, number of straight-chain carbon atoms), and the degree of physical blending with the resin mix.

The major types of lubricants or internal release agents are metallic stearates, fatty acids, fatty acid amides and esters, and hydrocarbon waxes.

Zinc stearate is the most widely accepted internal release agent in BMC and SMC. It has been reported that calcium stearate can be used at a lower concentration than zinc stearate and achieve the same result. The properties of zinc stearate that affect its use in BMC and SMC are particle size, absence of dusting, and residual alkali content.

TABLE 4.13 An Outline of Organic Peroxides

Recommended Mold Temperature	Vendor	Trade Names	Chemical Names	Form	Active Oxygen Content Percent	10 Hr. Half Life °C
180–230°F (82–93°C)	1	Percadox™F	Bis(4-t-butylcyclohexl peroxydicarbonate)	Powder		42
	2	Liladox™	Di-cetyl peroxydicarbonate	White powder	2.5	42
	3	Alperox™F	Lauroyl peroxide	Flake	3.9	62
230–260°F (110–127°C)	4	Vazo™64	Azobisisobutyronitrile	Crys.solid	DNA	64
	2	Lupersol™ 256	2,5-Dimethylhexane-2,5-diper 2 ethyl hexoate	Liquid	6.7	72
	2	USP-245 and Cadox™		Liquid	6.7	71
	6	Cadox™ BSP-50	Benzoyl peroxide in butyl benzyl phthalate	Paste	3.3	92
	1	Lupersol™231	1,1-Bis(t-butylperoxy)3,3,5-trimethyl cyclohexane	Liquid	10.58	
260–300°F (127–150°F)	2					
	6	Esperox™10	Tertiary butyl perbenzoate	Liquid	8.0	105
	2	TBPB	Tertiary butyl perbenzoate	Liquid	8.07	105
	1	Triganox™C	Tertiary butyl perbenzoate	Liquid	7.8	105
	5	DiCup 40C	Dicumyl peroxide	Powder	2.3	115
	2	Lupersol™101	2,5-Dimethyl-2,5-di(t-butyl peroxy) hexane	Liquid	9.9	119

Vendor Code:
1. Noury Chemical Corp.
2. Lucidol Div., Pennwalt Corp.
3. The Norac Co.
4. E. I. DuPont de Nemours & Co.
5. Hercules Powder Co.
6. U.S. Peroxygen Div., Witco Chem Co.

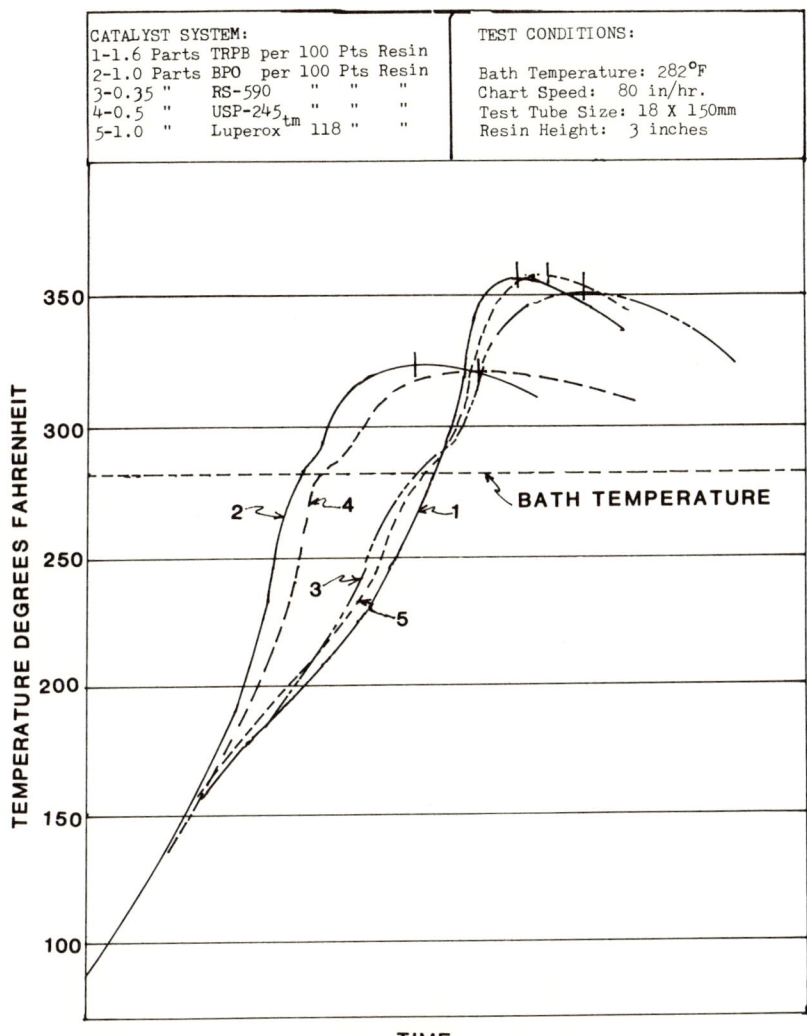

Figure 4.6A The Effect of Several Catalysts on the Cure Characteristics of a Resin Mix

Stearates are reported to have an effect on the initial thickening characteristics of SMC and, therefore, should be of consistent quality to eliminate variations in viscosity.

Table 4.14 (see p. 108) contains a listing of the melting points of several metallic stearates.

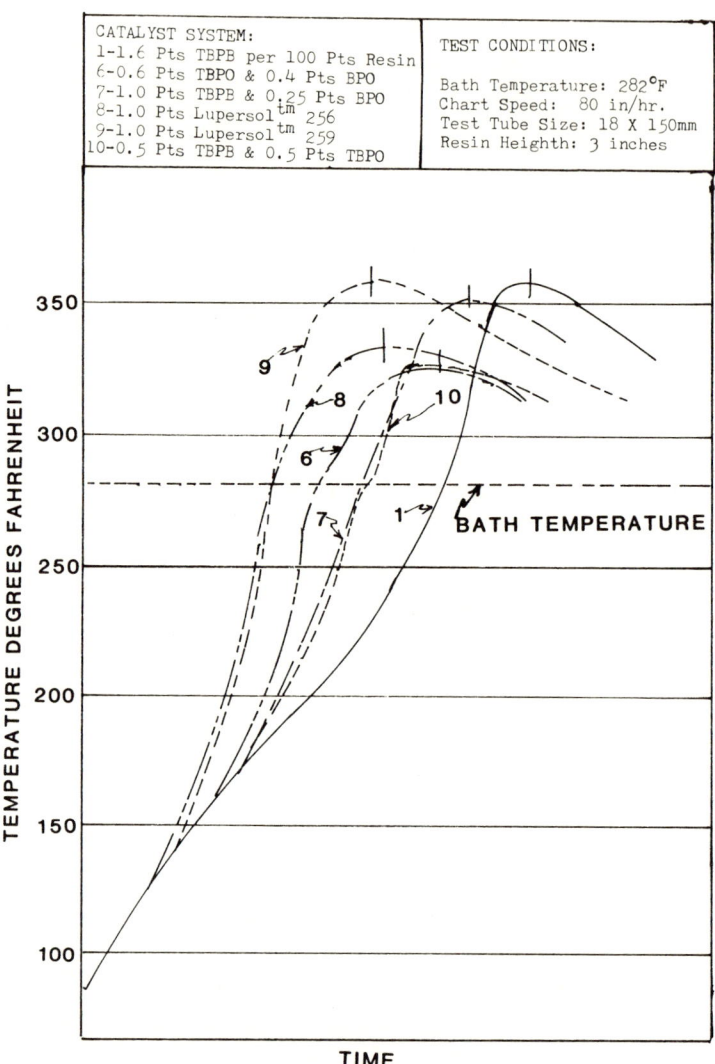

Figure 4.6B The Effect of Several Catalysts on the Cure Characteristics of a Resin Mix

Pigments

Pigments or colorants for BMC and SMC are becoming increasingly more important, as many products now are required to have color molded into the final part.

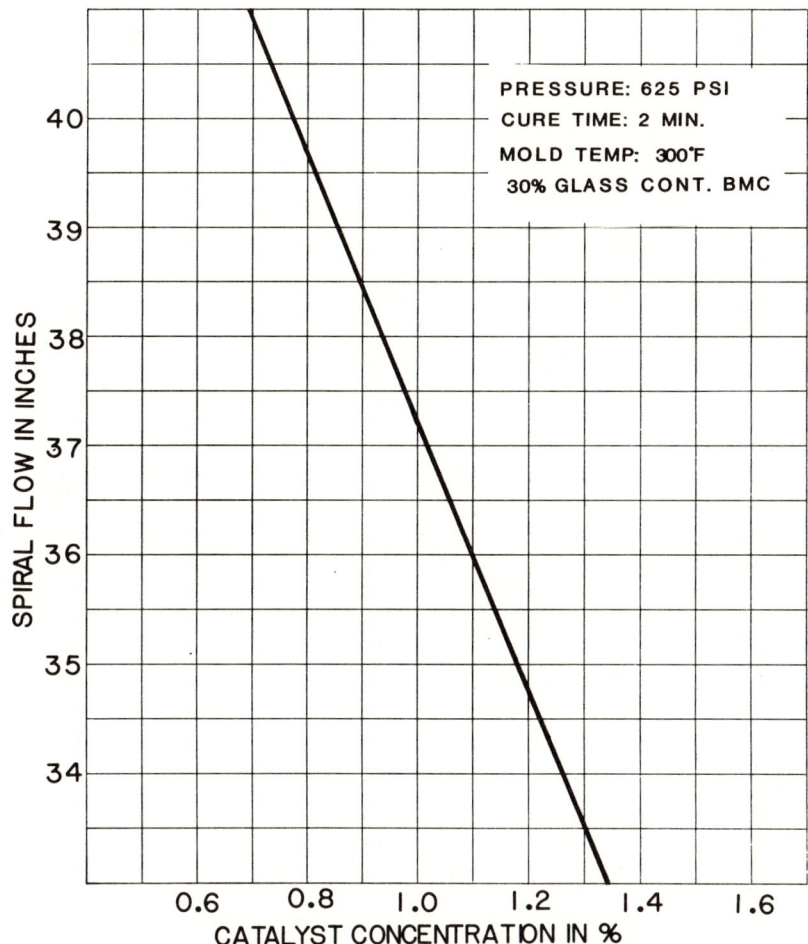

Figure 4.7 The Effect of Catalyst Concentration on Flow Properties of BMC

Colorants are divided into three basic families: dyes, organic pigments, and inorganic pigments.

Dyes are characterized by good transparency, high tinctorial strength, and low specific gravity. They have poor heat resistance and tend to migrate. For those reasons they seldom are used in polyester compounds.

TABLE 4.14 Melting Points of Several Metallic Stearates

Zinc Stearate	120–125°C
Calcium Stearate	150–160°C
Aluminum Distearate	145–160°C
Magnesium Stearate	154°C
Lithium Stearate	200–210°C

Organic pigments have good brightness and brilliance. They are divided into toners and lakes. Colors generally are not so bright as those obtained from dyes. Other characteristics are limited solubility, low specific gravity, high bulking value, and high oil absorption.

Inorganic pigments usually are natural or synthetic metallic oxides, sulfides, and other salts calcined at 1,200–2,100°F. Those pigments are superior to organic pigments and dyes in heat and light stability, weather resistance, and migration resistance. They are characterized by high specific gravity, low oil absorption, and good chemical resistance. In general that family produces opaque colors. Exceptions are the iron blue pigments that make a translucent shade in low concentrations.

In polyester resins a strong oxidizing action is present during polymerization. Good light-fastness and good dispersibility are required of the colorant as well as its being relatively neutral to polymerization. Some pigments have a strong effect either in accelerating or inhibiting the curing characteristics. Table 4.15 lists the effects of some pigments on the curing characteristics of polyesters.

In BMC and SMC dry powdered colorants sometimes are used (generally titanium dioxide, carbon black, or iron oxide), but most often a paste concentration is employed. Paste concentrations are color matched to prevent batch-to-batch variations. They are milled or ground into a vehicle. The dispersions vary in viscosity from a pourable to a heavy nonflowing paste, depending on the type of vehicle used, the pigment and solids content. In pastes, inhibitors and accelerators may be added to offset the effects of the pigments on polyester curing.

TABLE 4.15 The Effects of Pigments on the Cure Characteristics of Polyester Resins

Pigments	Effect
Cadmium Salts	Accelerators
Carbon Black	Inhibitor
Iron Oxide	Accelerator
Copper Salts	Inhibitor/Accelerator (Depending on concentration)
Aluminum Salts	Accelerators
Titanium Dioxide	Slight Accelerator
Phthalocyanine Green	Slight Inhibitor
Organic Dyes and Pigments	Inhibitors
Kaolin Clays	Slight Accelerators

Carrier Films

Two carrier films are generally used to transport the SMC product through the impregnator. Normally the carrier films are 3 to 4 inches wider than the SMC product to prevent resin mix from flowing out of the roll before chemical thickening has an opportunity to control the mix viscosity. Untreated extruded polyethylene film of 200 gauge (0.002 inches) is the most popular carrier film because of its more favorable cost. The carrier film should be pigmented in a color that will contrast with the SMC product to make it easier for the press operator, when preparing the mold charges, to be able to ensure that all the carrier film has been removed. Dow's type 505 film is an example of a satisfactory carrier film.

Polyethylene carrier film is fairly porous and will allow styrene monomer vapors to evaporate. For that reason, it is necessary to securely wrap the completed roll of SMC product with a barrier film, such as Mylar™ polyester film, aluminized kraft paper, cellophane, etc. Mylar™ works exceptionally well, but it is expensive and transparent, allowing light to penetrate the outer SMC layers. Aluminized kraft paper, a good barrier film, is opaque and fairly inexpensive.

One grade of uncoated cellophane has been tried as a carrier film, but it was partially solubilized by the styrene in the SMC resin mix and inhibited the cure of the SMC. When heated rolls are used on the impregnator, the cellophane tends to become brittle.

Polyethylene-coated cellophane has worked satisfactorily as an SMC carrier film, but it is not competitive with the extruded blown polyethylene film.

Nylon film as thin as 1 mil has worked exceptionally well as a carrier film and is superior to 2 mil polyethylene film when heated rolls are used, but it cannot compete economically with polyethylene.

All carrier films should be transported and stored horizontally, preferably with their tubes supported. Standing rolls of film on end causes the edges to distort, and that will put wrinkles on the damaged edges in use.

Chemical Thickeners

Many of the Group II oxides and hydroxides will thicken carboxy-terminated unsaturated polyester resins to some extent; however, the most popular thickeners are magnesium oxide, magnesium hydroxide, calcium oxide, calcium hydroxide, or combinations of those materials.

Magnesium oxide does not occur naturally. It is made by calcining magnesium salts. There are two types available commercially, which are produced by calcining either (1) mineral magnesite ($MgCO_3$) or brucite [$Mg(OH)_2$]; or (2) precipitated types obtained from seawater or natural brines.

Because of their relative nonreactivity the first type obtained from minerals is generally not used in polyesters. The type used to thicken polyesters is made by a controlled calcination of magnesium hydroxide, magnesium basic carbonate, or mixtures of these two products to give a porous, spongy, large surface area MgO. The surface area of a highly active MgO is of the order of 200 square meters/gram based on nitrogen absorption.

In general, increased thickening is directly related to higher surface area.

Besides reacting with polyester resins, MgO also reacts readily with water to form the hydroxide and with carbon dioxide to form the carbonate. In practical terms that means that the handling and storing of MgO is quite important in the SMC plant. For that reason some early SMC manufacturers arranged to purchase their MgO in small, sealed, moisture-impervious bags of 1 to 3 pounds each rather than in 50-pound bags. That special handling added considerably to the cost. Two companies actually make a business of packaging preweighed MgO, powdered pigments, etc., in polystyrene bags so that the entire bag can be tossed into the mix, where it dissolves without coming into contact with plant air.

A better solution has since been devised by the resin manufacturers, who have made available dispersions of reactive MgO in inert polymeric vehicles. One of the more popular such dispersions is Modifier "M" (USS Chemical Co.).

Magnesium hydroxide will also thicken some resin systems, and it can be used in combination with MgO. Calcium hydroxide can be an effective thickener, but it is less effective than MgO and it does not thicken all resin systems.

Additives

Toughener Additives

The highly reactive resins used in low profile BMC and SMC parts produce a fairly brittle product that sometimes is subject to profuse surface cracking during handling, shipping, and subsequent secondary machining operations. In an attempt to "toughen" those compounds to resist cracking, elastomeric additives are included in the formulation.

B.F. Goodrich Chemical Company has been promoting its Hycar™ Reactive Liquid Polymer (RLP) for several years as an elastomeric additive for rigid polyester systems. The recommended dosage is 10 phr (parts additive per hundred parts resin), and it is claimed that in calcium carbonate filled systems a considerable improvement in Gardner Impact Strength results.

The conjugated diene butyl (CDB) elastomer offered by Exxon Chemical Company is claimed to contribute low profile and antishrink behavior with improved physical properties. Optimum elastomer level is 7 to 9% based on total polymer/monomer content.

Some molders have combined a flexible polyester resin with a rigid isophthalic to achieve the same result as adding an elastomer and at considerably less cost. Several resin vendors now offer such a resin.

Viscosity Depressants

As inert fillers are added to a polyester resin, the mix viscosity increases rapidly. Certain materials act to depress the viscosity curve and are called viscosity depressants or viscosity control agents.

Calcium stearate acts as a viscosity depressant when china clays are added to a polyester resin system.

Union Carbide markets a material known as Y-9306 Dry Silane Concentrate, which acts as a viscosity depressant when alumina trihydrate is added to polyester resins. A loading of 4% concentrate, based on the alumina trihydrate content, is recommended.

Union Carbide also markets a product known as Viscosity Reducer-3 or VR-3, which is an effective viscosity control agent for calcium-carbonate-filled polyesters.

BYK-Mallinkrodt also markets surface active polymers that are effective in reducing the viscosity of inert filled polymers. Specific products are recommended for specific fillers. Generally a loading of 1% wetting agent is used based on the filler content.

BYK-Mallinkrodt 510 has been found to be an effective agent in eliminating air from highly loaded glass-reinforced BMC and HSMC. That agent allows better wetting of the glass, which results in higher physical properties, as compared with batches that do not contain the additive.

CHAPTER 5

Formulations & Compound Preparation

Introduction

This chapter deals with converting the raw materials into molding compounds. The order of ingredient addition to the mix is important, and the preferred order is given in Table 5.1 together with the terminology applied to the various phases of a mix.

For bulk molding compounds (BMC), when items 1 through 6 are combined, the resultant mix generally is referred to as a resin mix. Resin mixes should have long shelf lives. Resin mixes for BMC are normally charged to a double arm batch intensive mixer or are pumped to a continuous mixer where the filler is added to form a paste. If the mix is to be thickened, the chemical thickener is added at that point. The glass reinforcement is then added on a controlled mix cycle to complete the BMC. Regular unthickened mixes then are ready to be molded. For high glass content compounds it generally is desirable to age the compound at least one day to permit the glass reinforcement to more fully wet out in order to develop full physical properties. Thickened mixes must be aged in a temperature-controlled environment to complete the maturation to a predetermined molding viscosity. The length of that maturation period is determined by the nature of the polyester resin, the type and quantity of thickener used, and the temperature of the maturation room, which is normally maintained in the 90–105°F (32–41°C) range. The maturation period ranges from several days to several

113

TABLE 5.1 Molding Compound Raw Materials Order of Addition and Terminology

Bulk Molding Compounds		Sheet Molding Compounds		
Item Ingredient		Item Ingredient		
1. Polyester Resin		1. Polyester Resin		
2. Thermoplastic Additive		2. Thermoplastic Additive		
3. Styrene Monomer (if required)	Resin Mix	3. Styrene Monomer (if required)	Resin Mix	Paste
4. Organic Peroxide Catalyst		4. Organic Peroxide Catalyst		
5. Internal Release Agent		5. Internal Release Agent		
6. Pigments		6. Pigments		
7. Inert Fillers		7. Inert Fillers		
8. Chemical Thickeners (if required)		8. Chemical Thickeners		
9. Glass Reinforcement		9. Glass Reinforcement		

weeks. Most popular formulations are moldable in from 5 to 10 days.

For sheet molding compounds (SMC) the order of addition is the same as for BMC except that the inert filler is added directly to the resin mix in the mix tank. Generally a lower quantity of inert filler is used as compared with BMC. The inert filler is weighed and gradually charged to the mix tank as the agitator continues to rotate at a moderate speed (900–1200 RPM). Some systems use a hanging weigh hopper, which transfers the filler into the mix tank with an inclined U plane containing a vibrator. Others use a rotary auger to meter the weighed charge to the mix tank. The best results occur when the filler is continuously charged to the tank into the high side of the mix near an edge and not into the vortex of the mix. Adding filler into the vortex carries along a considerable amount of air. Many mix tanks contain an angle bolted or welded to the inside sides of the mix tank at 2 or 3 places to fold the rotating mix away from the sides (increase turbulence). On mixes that contain a high filler loading (more than 200 parts filler per 100 parts resin) it generally is necessary to raise the revolving blade several times during the filler addition to maintain the proper vortex. Mixing should be continued until the resin mix reaches a predetermined temperature, generally 100°F (38°C). That will require more mixing time during winter months than for summer periods.

For batch type SMC systems the chemical thickener should be added to the mix immediately before transferring the paste to the SMC impregnator. If the mix has been allowed to stand for an hour or more or if its temperature has decreased, the paste should be remixed until its temperature is 100°F (38°C) before the chemical thickener is added. Upon adding the thickener, the paste should be mixed until its temperature is 105°F (41°C) or for a minimum of 10 minutes.

Preparing Resin Mixes

A variable speed, high-shear, single-shaft mixer, usually referred to as a dissolver or dispersator, is excellent for preparing BMC and SMC resin mixes. Such a mixer is shown in action in Figure 5.1. The blade diameter should be approximately one-third the mix tank diameter for fastest dispersion. Turbine mixers and propeller type agitators also are used. Usually the polyester resin is weighed into the mix tank followed by the thermoplastic additive. Those two ingredients are agitated for a sufficient time to thoroughly disperse the thermoplastic in the polyester resin. The minor ingredients (styrene monomer, catalyst, release agent, pigments, and any other small-quantity additives) are weighed separately and charged to the mix tank while the blade rotates at a nominal speed (500–900 RPM). That prepared resin mix is then dumped into the double arm high intensity mixer or pumped to a continuous mixer.

A rotating horizontal shaft mixer containing plows inside a drum (made by Day Mixing Co.) is sometimes used to prepare BMC and SMC pastes. On some models the BMC can be completely formulated in this mixer by adding the glass reinforcing strands directly to the mixer at the end of the mix cycle. Figure 5.2 is such a mixer.

Figure 5.1 Typical Dissolver-Type Mixer Preparing a Resin Mix

Table 5.2 is a troubleshooting guide that should be useful in overcoming problems when SMC is being prepared.

Resin Mix Dispensing Systems

One of the first dispensing systems for SMC pastes consisted of a hold tank (50–100 gallons) and a Grayco or Alemite air operated pump to transfer the paste to the doctor boxes on the SMC impregnator. A tee was placed in the supply line to divert the stream to the two doctor box stations with a ball valve on each side of the tee to control the paste flow. That system

Formulations & Compound Preparation 117

Figure 5.2 Littleford Mixer Capable of Preparing Resin Mixes and BMC (*Courtesy: Day Mixing Company*)

worked very well on the early SMC pastes, which did not reach a viscosity more than 100,000 centipoises in the first two hours after adding the chemical thickener. Some laboratories still use this sytem of dispensing. Even with those systems on the early mixes a variation in glass wet out frequently was observed from the start of a batch to the end of the batch of paste.

As resins and chemical thickeners changed to more quicker maturation systems, the variation in glass wet out from start to finish of a batch became intolerable and a considerable amount of development work was expended to improve paste dispensing systems. The idea of the 2-pot system evolved using the A side or resin side and the B side or thickener side. The A side contains all of the polyester resin, thermoplastic low profile additive, the catalyst, release agent, and the majority of the inert filler. The B side consists of an unthickenable carrier resin, the pigment, chemical thickener, and the balance of the inert filler.

TABLE 5.2 Troubleshooting SMC Processing Problems

Problem	Description	Possible Cause	Remedy
Film Wrinkles	Film wrinkles at doctor box station	Low film tension	Increase film tension
		Misaligned slat expander	Readjust the slat expander
		Insufficient carrier film to expander roll contact	Reposition the slat expander to increase the carrier film contact
		Roll of film out of alignment with film travel	Realign the carrier film roll
		Film drag on the doctor box platens	Clean the plates to remove any contamination
Film Tears	At the edges of the doctor boxes	Insufficient clearance between side dams and film	Raise the side dams
		Edge damage on the film roll	Replace the roll of film
	In paste deposit area of the doctor blades	Filler lumps or contamination in the paste	Strain the paste
Streaky Paste Deposit	Dry lines in paste parallel to film travel	Filler lumps or contamination in the paste	Remix and strain the paste
Dry Fibers	Uneven (thick thin) sections in SMC	Fiber clumps dropping from cutter or framework of cutter	Prevent fiber glass accumulations on ledges and static bars. Install air jets
		Resin-starved and/or glass-rich SMC	Adjust resin mix feed. Adjust side dams on roving chopper
		Too high viscosity	Control the humidity in mix room and in SMC room. Check mix ingredients for both proper type

Formulations & Compound Preparation

Problem	Cause	Remedy
	Insufficient number of compaction rolls	and concentration. Verify by assay if necessary. Check thixotropy of SMC soups and pastes. Verify water content of major mix ingredients. Install additional compaction rolls.
	Premature chemical thickening	Control the humidity in mix room and SMC room. Check mix ingredients for both proper type and concentration. Verify water content of all major ingredients
	Eddy currents in the glass deposition area	Enclose the chopper area. Remove any fans from the area.
	Static accumulations	Install static bars. Ground all framework and rolls
Nonuniform Fiber Distribution	Uneven roving spacing	Restring all rovings
	Insufficient number of rovings	Add additional rovings to creel
	Static accumulation	Install static eliminators. Ground framework and rolls
	Thick and thin areas across SMC width	
	Roving patterns into chopper too wide	Narrow roving pattern at cutter comb
	Side shields improperly adjusted	Readjust chopper side shields
	High fiber content near SMC edges	
	Air currents under the chopper	Enclose hopper frame to eliminate air currents
	Roving pattern too narrow at chopper	Widen roving pattern at cutter comb
	Side shields improperly adjusted	Readjust side shields on chopper. Adjust roving pattern at chopper comb
	Low fiber content near SMC edges	

TABLE 5.2 Continued

Problem	Description	Possible Cause	Remedy
Long fibers	SMC otherwise appears normal	Uneven roving pattern into chopper Chopper operating improperly	Check cutter cot and roll alignment Replace cutter blades Replace worn cutter cot Increase pressure on cutter cot
Fiber Build-up on Cutter Frame/Rolls	Fiber clumps on framework on cutter roll, and/or on cot	Excessive static Abrasion of rovings at creel, guides, tubes, etc. Low humidity in area Roving pull speed too high	Check static eliminators for proper operation. Replace damaged pins if necessary Check for roughness at all roving contact points Increase humidity Increase number of rovings to chopper and reduce chopper speed
Air Between Fibers And SMC		Insufficient compaction roll pressure Resin-rich SMC edges	Increase roll pressure Adjusted for small amount of dry edge glass
Poor Fiber Wet-Out	Random distribution	Uneven glass blanket Paste viscosity too high Insufficient compaction pressure	Check for static and chopper condition Reduce paste viscosity Increase compaction roll pressures

Formulations & Compound Preparation

Problem	Possible Cause	Corrective Action
Poor Fiber Wet-Out	Glass content too high for machine capabilities	Increase roll temperatures. Change glass to one more easily wet-out
Localized lane of SMC	Improper spacing of rovings into the chopper Nonuniform paste deposit on films Nonuniform compaction pressure	Respace the rovings in the comb Check doctor blade set-up. Check for contamination on back-up plate Check compaction rolls for distortion Check for contamination build-up on compaction rolls
Squeeze-out At Winding Station	Excessive winding tension Paste viscosity too low for glass content being used Film too narrow for SMC width	Reduce winding tension Adjust paste viscosity Get wider film or reduce SMC width
Telescoping of SMC at Wind-up Station	Rolls are egg shaped Excessive winding tension Poor alignment of winder with SMC machine	Reduce winder with the machine Realign the winder with the machine
Improper Weight Per Unit Area	Sheet appears normal and uniform Glass and/or resin quantities need adjustment Incorrect machine calibration Resin mix ingredients incorrectly proportioned	Check glass content. Increase glass content (raise cutter speed) and/or paste feed (raise doctor blade) as required Recalibration machine settings Check specific gravity of resin mix

One of the first dispensing systems to use the A and B side concept consisted of two connected double acting cylinders with adjustable pistons mounted on a single shaft. Air-operated solonoid valves were used on the cylinder ports. The driving force for moving the piston was furnished by the pressure of the resin mixes, which were pumped to the cylinders. A limit switch interrupted the piston travel and reversed the travel direction so that the exit ports became the entrance ports and vice versa. At one installation, two holding tanks were used for the A and B sides, and gear pumps were used to move the two resin mixes through the cylinders. Another used pressure pots with a nitrogen charge over the two resin mixes as the moving force. Both molders used an air-driven piston pump to move the resin mixes from the mix house to the holding tanks. While that system did permit continuous operation of the SMC impregnator, it was not without its problems. Even though the seals on the cylinders were Teflon™, they wore out easily from the abrasive mix and had to be replaced weekly. The connecting lines were of small diameter, and solonoid valves tended to stick. Maintenance costs were high and down time frequent.

In its laboratory the Koppers Company perfected a dispensing system, which used a Moyno™ pump on the A side and a gear pump on the B side. It was a considerable improvement over previous systems.

The Marco Chemical Division of W. R. Grace & Co., after considerable effort and many changes, built a similar system but used Moyno™ pumps of different sizes for both the A and B sides. Some of the systems mentioned above are now commercially available from Finn & Fram, E. B. Blue Co., and others.

Static Mixers

In using the 2-pot system the A and B mixes must be combined before they are fed to the doctor box stations. That combination can be accomplished by the use of a static mixer. The literature describes the operation of such a unit as achieving a homogeneous mix by flowing the two streams through geometric patterns formed by elements mounted in a tubular barrel. It has been found that a stainless steel mixer of 1-inch pipe size, with a minimum of 21 elements and preferably 24 elements, will do a reasonable job with SMC resin mixes. The static mixer should be installed in such a manner that the elements are easily removable for periodic cleaning in methylene chloride. Adding all the pigment to the B side resin mix allows one to visually judge the degree of mixing at the doctor boxes. A homogeneous mix without streaks is required.

Dynamic Mixers

While static mixers do a good job in combining the two resin mix sides, when time is available for periodic thorough cleaning of the mix lines and elements, they leave much to be desired for continuous operations 24 hours a day seven days a week. For such operations a dynamic in-line mixer is required. One such mixer that has proven to be satisfactory on SMC impregnators is made by Chemineer Inc.

In-Process Testing of SMC

SMC Batch Process

The following are the minimum in-process tests that are recommended on an SMC batch process to keep it under control:

1. Weight per unit area
2. Glass content
3. Glass wet out
4. Initial paste viscosity
5. Final paste viscosity
6. Critical process temperature monitoring

SMC Continuous Process

The following are the minimum in-process tests that are recommended for a continuous or semicontinuous SMC process to keep it under control:

1. Weight per unit area
2. Glass content
3. Glass wet out
4. Initial paste viscosity
5. Final paste viscosity
6. Process temperature monitoring
7. Ratio check of A and B streams

Test Frequency

1. Weight per unit area check: At impregnator start-up. Once each hour thereafter. After each process adjustment.
2. Glass content determination: At impregnator start-up. Once per hour thereafter. After each process adjustment.
3. Glass wet-out check: At impregnator start-up. Once per hour thereafter. Whenever weight per unit area, glass content, or viscosity checks are high. After a process adjustment or a new lot of glass roving has been added to the line.
4. Initial paste viscosity measurement: At impregnator start-up. Whenever a new lot of paste is charged to the line.
5. Final paste viscosity measurement: After 1-hour aging of the paste sample. After 24-hour paste aging. Daily thereafter until all the SMC has been molded.
6. Critical process temperature monitoring: Continuous recording of all critical process temperatures are recommended.
7. Ratio checks on A and B sides: At impregnator start-up. Once per hour thereafter. Whenever a new batch of A side resin mix or B side thickener is charged to the line.

Test Procedures

1. Weight per unit area check. English Units: The SMC is placed on a flat surface and a 12 × 12 inch 12-gauge steel template with a handle is placed on top of the SMC. With a knife the SMC is carefully cut along the template edges as it is firmly held in place. The sample is weighed to the nearest 0.1 gram.

$$\text{Calculations:} \frac{(\text{sample wt}) - (\text{PE film wt})}{28.3}$$
$$= \text{Weight in ounces per square foot.}$$

For 24-inch wide sheets two samples are taken across the width of the sheet. Those samples can be identified as left and right according to their orientation to the machine direction. For sheets 36 inches and wider samples are taken across the width of the sheet and identified so as to relate to their orientation in the sheet.

2. Glass content check. This test usually is performed on the samples of SMC previously cut out for weight per unit area checks. A 6 × 6 inch (150 × 150 mm) sample is cut from each of the area weight samples and weighed to the nearest 0.1 gram. Each sample is placed in an individual solvent tank containing methylene chloride mounted in an ultrasonic bath. After the tank has been vibrated a few minutes, the pieces of polyethylene film may be retrieved with a metal tong, rinsed with fresh methylene chloride from a wash bottle, dried, and the film weight subtracted from the original SMC sample weight. The sample is then vibrated in the bath until all the filler has been removed. The glass fiber is removed, washed with fresh methylene chloride, dried, and the glass is weighed to the nearest 0.1 gram.

$$\text{Calculations: \% Glass} = \frac{\text{Fiber weight}}{\text{Net SMC weight}} \times 100$$

3. Glass wet-out check. The degree of wetting of the glass reinforcing fibers by the resin mix is an extremely important

property of all SMC sheets, but it is a very difficult property to completely define in the laboratory or plant. One method consists of preparing standards of various degrees of wetting and of making full-scale photographs of those standards to be used by the technician in grading SMC samples. Numbers are assigned to the various standards, starting with zero for a blanket of dry glass, with the numbers increasing as the degree of wetting increases until 10 is assigned to a film of resin without glass fibers. It is not necessary to have comparison standards for each number in the series. Interpolation between numbers is fairly simple to do. Samples for the procedure are obtained from the pieces left over after glass content samples have been removed from the unit area samples. The film is stripped off one side, and the appearance of the sample is compared with the photograph standards. A number is assigned to the degree of wetting the sample is judged to be. When the sample is judged to be an 8 or higher, the machine is allowed to run uninterrupted. If the sample is judged to be a 7, extra samples are cut at the next complete roll and additional inspections made. If the rating continues to be 7 on the next check, remedial action must be taken.

4. Initial paste viscosity. Two pint samples of paste are taken of each batch of paste toward the center of the batch run. One sample is stored with the SMC; the other is taken to the Quality Control laboratory, where its temperature and Brookfield viscosity are measured and recorded.

5. Final paste viscosity. The initial viscosity sample is maintained at the same temperature as the SMC (preferably in a water bath), and its viscosity is again measured after it has aged for 4 hours and after 24 hours. After each test the sample is returned to its controlled temperature environment. The procedure should be repeated every 24 hours until the SMC has been molded. The sample of paste stored with the SMC can be used to verify the viscosity of the laboratory-stored sample.

In-Process Testing of BMC

Usually the Brookfield viscosity and SPI cure characteristics of the BMC resin mix are checked on each batch prepared. After the inert filler is added, the mix is usually not checked. Tests on the completed compound will be found in Chapter 7.

BMC Formulations

PPG Two-component and Three-component Systems

One of the better premix studies published by raw materials vendors was published by PPG Industries in 1957 and was revised and reissued in 1958.[45] The report is the result of several designed experiments intended to determine the effects of sisal, asbestos, and glass reinforcement combinations on the physical properties, appearance, moldability, and handleability of premix compounds. At the time the report was completed, glass fibers were selling for 40 cents per pound, sisal for 20 cents, and asbestos fibers for 2.1 cents per pound. All of those materials have increased dramatically in price and not all at the same rate. Sisal no longer has the big advantage in cost over glass fibers and is very seldom used today. Asbestos was used to dry up the mix, but since it now is a suspected carcinogen, it has been replaced by other fibrous fillers, such as wollastonite. The report, however, does contain useful data on all glass compounds at 15, 20, and 25% loadings by weight and on several filler levels.

OCF Report

In 1969, Frankenhoff, Maxel, and Miller[166] published six different types of formulations (general purpose, self-extinguishing, electrical, corrosion resistant, and two separate appearance grades), each at three different glass contents together with physical properties and costs. The cost data are out of date, but everything else still applies.

Published Formulations

The following formulations were taken from the published literature and from vendor's bulletins.
Formulation Number: BMC-1
Source: British Plastics, (Sept. 1970)
Reference: Cleaver[205]
Type Compound: Low Profile Injection Moldable BMC

A. Formulation

	As Published	% By Weight
1. Paraplex™ P-19-A	35%	33.45
2. t-butyl perbenzoate	1% BOR	0.97
3. Zelec™ UN	2% BOR	1.91
4. Triton™ GR-7	0.5% BOR	0.47
5. Dacote™	45%	43.16
6. Microthene™ FN-510	1% BOR	0.97
7. Glass; OCF-832-¼ in. lengths	20%	19.07
8. Total Compound		100.00

B. Raw Materials notes

1. 60% P-340 & 40% P-543; Formerly Made By Rohm & Haas Co., Now by OCF.
3. E. I. duPont de Nemours & Co.
4. Wetting agent made by Rohm & Haas Co.
5. Diamond Shamrock coated calcium carbonate (no longer available)
6. U.S.I. Chemical Co.

C. Physical Properties

1. Flexural Strength, psi	15,300
2. Flexural Modulus, psi × 10^6	1.57
3. Izod Impact Strength, ft. lbs./in. notch	5.3

Note: BOR = based on resin

Formulation Number: BMC-2
Source: Amoco Chemicals Corp.
Reference: Bulletin 1P-31b[255] (Feb. 1972)
Type Compound: Low Profile BMC

A. Formulation

	Percent By Weight
1. Resin SG-30 (30% styrene monomer)	13.9
2. Low-Profile Additive (40% solids)	9.3
3. t-butyl Perbenzoate	0.18
4. Zinc Stearate	0.5
5. Calcium Hydroxide	0.23
6. Calcium Carbonate	60.7
7. Glass, ¼ in. lengths	15.1

B. Raw Materials Notes

1. Resin formulation also provided in Bulletin 1P-31b
2. Amoco M9-C2 polystyrene in styrene monomer
4. Parson Plymouth Grade XXX-H
6. Diamond Shamrock "Suspenso"
   (no longer available)
7. Johns-Manville's CS-308AG
   (no longer available)

C. Physical Properties

1. Flexural Strength, psi	14,900
2. Flexural Modulus, psi × 10^6	1.79
3. Izod Impact Strength, ft. lbs.; inch notch	6.4
4. Tensile Strength, psi	6,400
5. Barcol Hardness	60
6. Water Absorption, percent	0.09

Formulation Number: BMC-3
Source: Private communication; Hooker Chemical Co. (now Occidental Petroleum)
Reference: Gouinlock and Smith, "Studies on the Processability of SMC and BMC," Mar. 15, 1973 (previously unpublished).
Type Compound: Chemically thickened low profile BMC.

A. Formulation

1. Durez 29,261 One Component Polyester Resin	100
2. Catalyst	varies
3. Zinc Stearate	3
4. Magnesium hydroxide	1.5
5. Calcium Carbonate	200
6. Glass Fibers, ¼ in. lengths	75
7. Total Compound	379.5+

B. Raw Materials Notes

1. Occidental Petroleum no longer markets polyester resins, having sold its polyester business to Ashland Chemical Co.

C. Physical Properties

Some of the data generated in this instrumental mold study are shown in chart form, as follows:

Figure 5.3: Effect of Mold Temperature and Catalyst Type on Rib Profile
Figure 5.4: Effect of Mold Temperature on BMC Cure Time
Figure 5.5: Effect of Mold Temperature on Flexural Strength of BMC
Figure 5.6: Effect of Mold Temperature and Catalyst Type on BMC Shrinkage

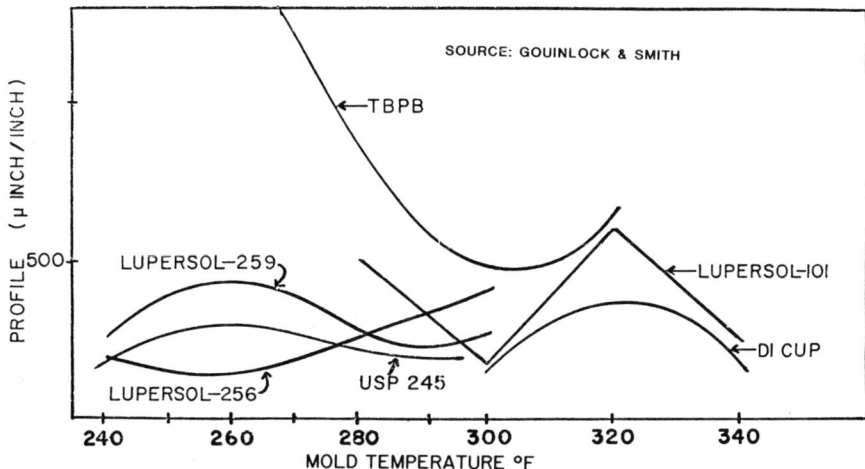

Figure 5.3 Effect of Mold Temperature and Catalyst Type on Rib Profile (*Source: Gouinlock & Smith*)

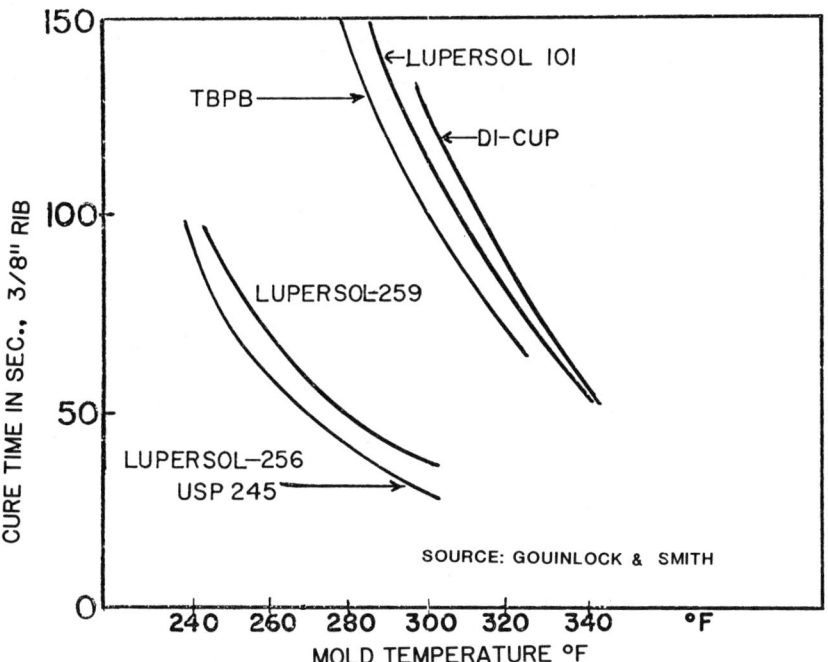

Figure 5.4 Effect of Mold Temperature on BMC Cure Time (*Source: Gouinlock & Smith*)

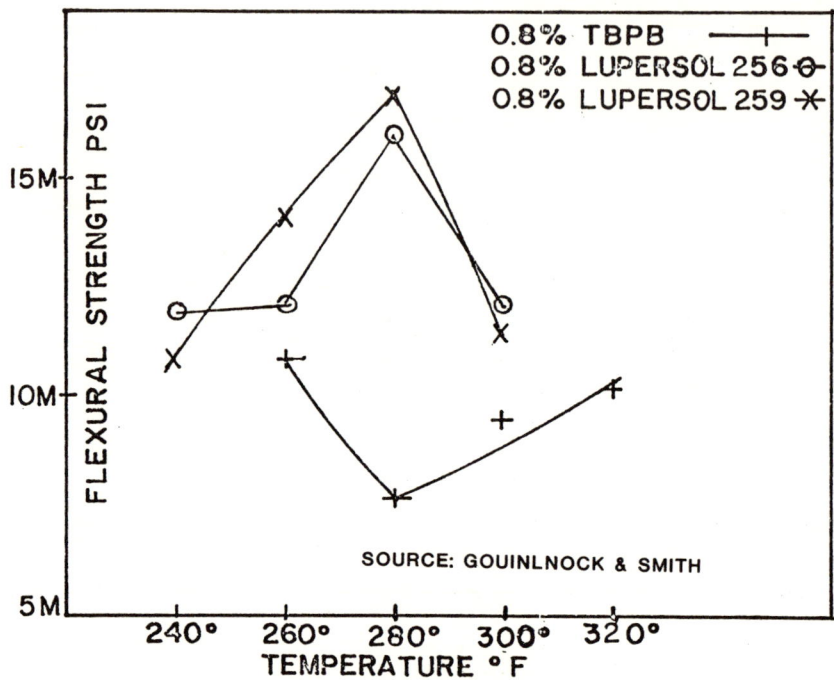

Figure 5.5 Effect of Mold Temperature on Flexural Strength of BMC (*Source: Gouinlock & Smith*)

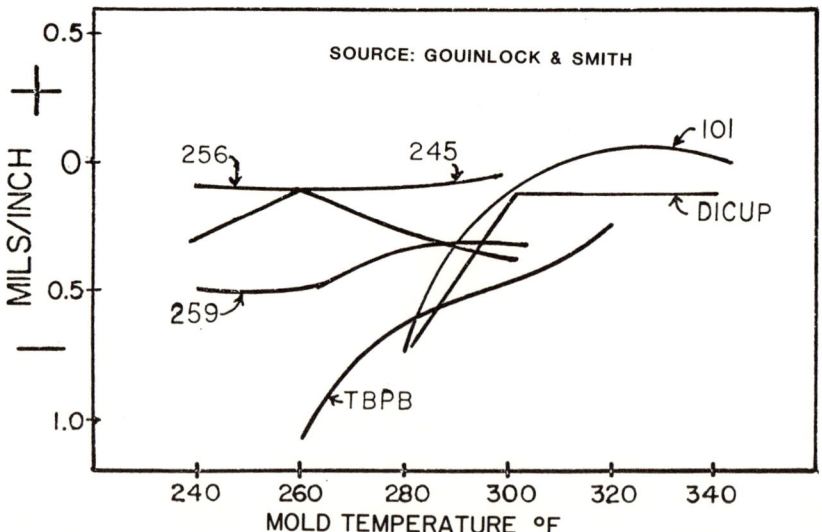

Figure 5.6 Effect of Mold Temperature and Catalyst Type on BMC Shrinkage (*Source: Gouinlock & Smith*)

Formulation Number

		BMC-4	BMC-5	BMC-6
1.	Koppers 3700-25 Resin	100	100	100
2.	Styrene Monomer	9.1	9.1	0
3.	Thermoplastic Resin L	65.7	28	0
4.	5% PBQ Solution	1.1	1.1	1.1
5.	Microthene 500	0	0	15
6.	t-butyl Peroctate	0.91	0.69	0.49
7.	Zinc Stearate	7	5.6	3.78
8.	Suspenso	420	322	202
8.	Calcium Oxide	.056	0.28	0.28
9.	Calcium Hydroxide	1.12	0.42	0.56
10.	Glass, ¼ OCF-832	107	81.2	56.6

B. Raw Materials Notes

3. A Thermoplastic Syrup Available From Koppers Co.
4. 5% p-benzoquinone in Dibutyl Phthalate
5. U.S.I. Chemicals
8. Diamond Shamrock (no longer available)
9. J. T. Baker Reagent Grade
10. J. T. Baker Reagent Grade

C. Physical Properties

1.	Barcol Hardness	62–65	62–65	45–50
2.	Flexural Strength, psi @ 77°F	13,000	14,000	16,000
3.	Flexural Modulus, psi $\times$ 10^6	1.6	1.6	1.2
4.	Flexural Strength, psi @ 175°F	11,400	12,500	15,000
5.	Tensile Strength, psi	5,900	5,300	7,300
6.	Izod Impact, Unnotched	5.5	6	8.5
7.	Glass Content, %	15	15	15
8.	Surface smoothness, micro in.	170	280	125

Formulation Number: BMC-7
Source: Paper 2-A, Proceedings SPI RP/CI 1980 meeting, New Orleans
Reference: Horner, Zaske, and Husain[529]
Type Compound: Low profile injection moldable BMC

A. Formulation

1. DL-208 Resin	27.5
2. Triganox™ 29-B75 Catalyst	0.5
3. Pigment	0.3
4. Release Agent	1.7
5. Camel-White Calcium Carbonate	50
6. Glass Fibers, ½ in. length	24
7. Total Compound	100

B. Raw Materials Notes

1. Isophthalic Polyester Resin by Silmar Div., Vistron Corp.
2. Noury Chemical Co.
5. Harry T. Campbell Co.

C. Physical Properties

1. Flexural Strength, psi (crosswise to flow)	11,600
2. Flexural Strength, psi (parallel to flow)	8,030
3. Flexural Modulus, psi × 10^6 (crosswise)	1.40
4. Flexural Modulus, psi × 10^6 (parallel)	1.25
5. Izod Impact, ft. lbs./inch (crosswise)	3.95
6. Izod Impact, ft. lbs./inch (parallel)	2.36

Formulation Number: BMC-8
Source: Lubin, G., ed., *Handbook Of Composites*, chap. 15, "Thermoset Matched Die Molding" by P. Robert Young, pp. 420–21, Van Nostrand Reinhold, New York, 1980.
Designation: FR halogenated resin premix

Formulation

Percent by Weight
1. Halogenated Resin   34
2. Catalyst, Benzoyl Peroxide Paste   0.6
3. Zinc Stearate   1.4
4. China Clay   44
5. ¼ in. Glass Fibers   20

Formulation Number:
Source: Thompson, Weinman Technical Bulletin, "Optimum Calcium Carbonate Selection for Reduced Density in SMC/BMC," undated.

A. Formulation

	BMC-9	BMC-10	BMC-11
1. GR-13024 Polyester	100	100	100
2. Zinc Stearate	3	3	3
3. T-butyl Perbenzoate	1	1	1
4. B37/2000 glass bubbles	0	7.5	15
5. Snowflake P.E.	220	165	110
6. Glass Fibers, ¼ in.	80	80	80

B. Physical Properties

1. Specific Gravity	1.95	1.75	1.53
2. Flexural Strength, psi	21,339	17,786	14,158
3. Tensile Strength, psi	6,849	5,617	4,206
4. Notched Izod	7.0	6.5	6.3
5. Barcol Hardness	60	50	32

Source: Thompson Weinman Co. Technical Bulletin, "Optimum Calcium Carbonate Selection for Reduced Density in SMC/BMC"

A. Formulation

Formulation Number:	BMC-12	BMC-13	BMC-14
1. GR-13024	100	100	100
2. Zinc Stearate	3	3	3
3. T-butyl Perbenzoate	1	1	1
4. Glass Bubbles B37/2000	0	7.5	10
5. Atomite	220	165	165
6. Glass Fiber, ¼ in.	80	80	85

B. Physical Properties

1. Specific Gravity	1.97	1.78	1.70
3. Flexural Strength, psi	18,748	19,480	20,308
4. Tensile Strength, psi	6,363	6,521	7,060
5. Notched Izod	6.0	6.9	7.4
6. Barcol Hardness	62	52	56

Reference: Stoops & Maxel[188]

A. Formulation

	Parts By Weight	Percent
1. Isophthalic Resin	100	31.08
2. Styrene Monomer	1	0.31
3. Microthene FN-510	15	4.16
4. Zinc Stearate	5	1.56
5. Organic Peroxide Catalyst	1	0.31
6. Calcium Carbonate	100	31.08
7. Calcium Hydroxide	3.24	1.00
8. 1½ in. Chopped Strands	96.5	30.00
9. Total SMC	321.77	100.00

B. Raw Materials Notes

3. USI Chemicals Co.

C. Molding Conditions

1. Cure time, 100 mils, min.	2.5
2. Molding pressure, psi	900
3. Mold temperature,	300

D. Mechanical Properties

1. Density, #/cu. in.	0.063
2. Tensile Strength, psi	12,000
3. Flexural Strength, psi	28,000
4. Flexural Modulus, psi	1,600,000
5. Izod Impact Strength, ft. lbs./in.	12
6. Reverse Impact Strength, in.	5

Formulation Number: SMC-2
Source: PPG Industries Technical Bulletin "SMC-A Systems Approach"
Designation: Selectron™ 50,239/5990 SMC
Type Compound: Automotive grade SMC

A. Formulation

	Parts	Percent
1. Selectron™ 50,239	60	16.13
2. Selectron™ 5990	37	9.94
3. Styrene Monomer	3	0.81
4. T-Butyl Perbenzoate	1.5	0.40
5. Zinc Stearate	2.0	0.54
6. Calcium Carbonate	150	40.31
7. Magnesium Hydroxide	3.25	0.87
8. PPG-518 Roving	115.35	31.00
9. Total SMC	372.10	100.00

B. Properties of Liquid Resin

1. Type resin; High Exotherm IPA Polyester. Note: Some PPG polyesters were sold to Ashland Chemicals Co.

2. Specific Gravity:	1.11–1.13
3. Viscosity in Centipoises	1600–1900
4. Acid Number	22–27
5. Water Content, percent	0.1 max.
6. SPI Gel Time in 250°F Bath	1–2 min.
7. SPI Time to Peak Temperature	2–3
8. SPI Peak Temperature	460°F min.

C. Average Physical Constants of Raw Materials

Material Property	Paraplex™ P-340	Paraplex™ P-681
Brookfield Viscosity, cps.	1,040	1,280
Specific Gravity	1.113	0.980
Acid Number	20	
Water Content, %	0.10	
Solids Content, %		32.5

D. Average Physical Properties of P-19C SMC (@ 30%-1 Glass)

Flexural Strength, psi	26,500
Flexural Modulus, psi	1,900,000
Tensile Strength, psi	13,500
Tensile Modulus, psi	1,850,000
Izod Impact Strength, Notched, Ft. Lb./In.	13
Barcol Hardness	60–65
Water Absorption	0.4–0.8
Specific Gravity	1.85

Formulation Number: SMC-12
Source: Rohm & Haas Technical Bulletin
Source: Designation: P-19C 2-Pot SMC
Reference: Rohm & Haas Technical Bulletin
Type Compound: Low-Profile SMC

A. Formulation

B. General Notes

1. This is an example of a useful combination where equal volumes of the two sides are combined. Each side contains approximately equivalent viscosities. There is considerable flexibility in devising such combinations as long as the final paste contains the proper concentrations of each ingredient.

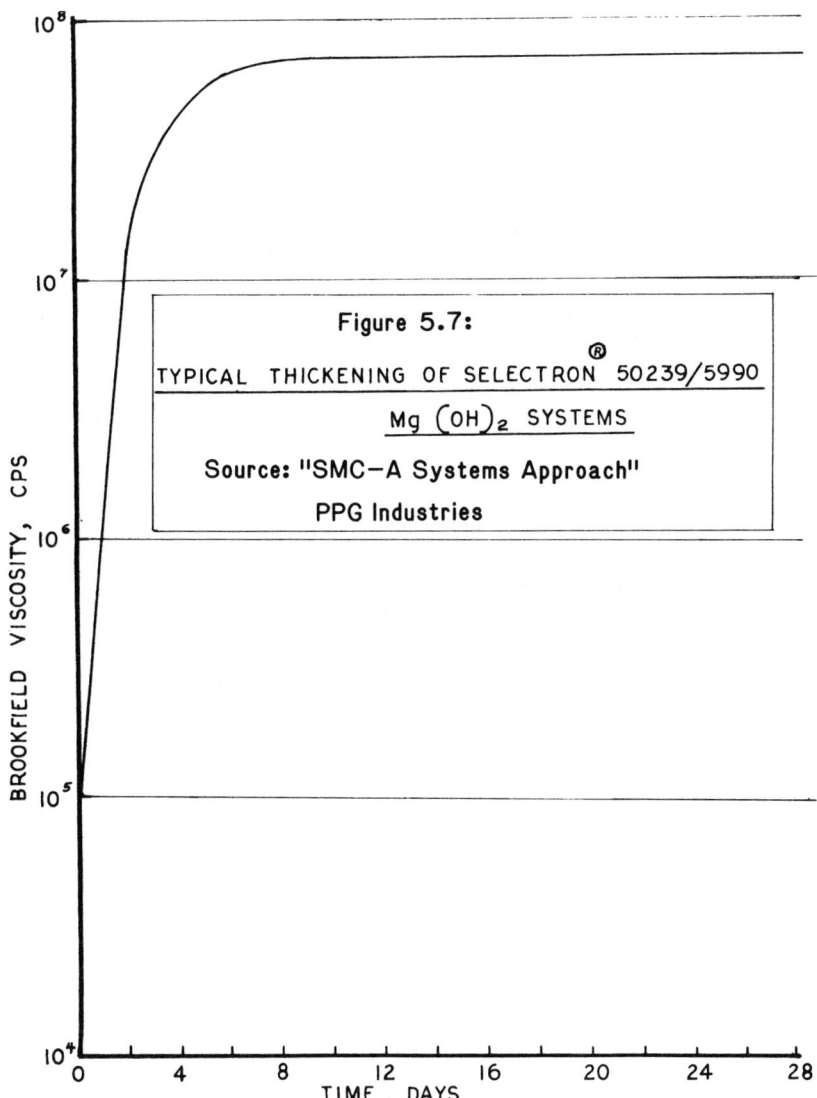

Figure 5.7 Typical Thickening of Selectron™ 50239/5990 Mg(OH)$_2$ Systems (*Source: SMC—A Systems Approach" PPG Industries*)

2. No more than 25% of the Paraplex™ P-681 should be replaced by the Paraplex™ P-543.
3. Neither Paraplex™ P-340 nor Paraplex™ P-681 is suitable as carriers for the Thickener.
4. Paraplex™ polyester resins formerly made by Rohm & Haas Co. are now made by Owens-Corning Fiberglas Corp.

Formulation Number: SMC-13
Source: Rapp, R. S., "High Strength Molding Compounds" (Oct. 1976), SPI RP/CI Thermostat Press Molding Committee meeting.
Type Compound: High Strength SMC

A. Formulation

	PHR	Percent
1. SG-30 Resin @ 60% solids	100	27.32
2. Zinc Stearate	3	0.82
3. t-Butyl Perbenzoate	1	0.27
4. MgO Dispersion	6	1.64
5. Glass, 1" PPG-518	256	69.95
6. Total Compound	366	100.00

B. Compound Properties

1. Masterbatch Viscosity	600–800 cps
2. 1-hour Thickened Viscosity	800–1400 cps
3. Time to 20×10^6 Viscosity	48–72 hrs.
4. Sheet Weight	7 oz./sq. ft.

C. Typical Physical Properties

1. Flexural Strength, psi	61,000
2. Flexural Modulus, 10^6 psi	2.50
3. Tensile Strength, psi	30,000
4. Tensile Modulus, 10^6 psi	2.40
5. Izod Impact, notched ft. lb./in.	21.5
6. Izod Impact, Unnotched, ft. lb./in.	28.6
7. Shrinkage, mils/in.	1

Formulation Number: SMC-14
Source: Rapp, R.S., "High Strength Molding Compounds" (Oct. 1976), SPI RP/CI Thermostat Press Molding Committee meeting.
Type Compound: Filled High-Strength SMC

A. Formulation

1. SH-30 Resin @ 60% Solids/Styrene	100
2. Atomite Calcium Carbonate	40
3. Zinc Stearate	3
4. t-Butyl Perbenzoate	1
5. MgO Dispersion	6
6. Glass, 1 in. Lengths	217

B. Physical Properties

1. Flexural Strength, 10^3 psi	54.8
2. Flexural Modulus, 10^6 psi	2.54
3. Tensile Strength, 10^3 psi	30.5
4. Tensile Modulus, 10^6 psi	2.83
5. Izod Impact Strength, ft. lb./in. not.	23.3
6. Izod Impact, Unnotched, ft. lb./in.	31.0
7. Shrinkage, mil/in.	0.37

Formulation Number: SMC-15
Source: Rapp, R. S., "High Strength Molding Compounds" (Oct. 1976), SPI RP/CI Thermostat Press Molding Committee meeting.
Type Compound: Low-Profile High-Strength SMC

A. Formulation

1. SG-30 Resin	85
2. LP-40A Thermoplastic	15
3. Atomite Calcium Carbonate	20
4. Zinc Stearate	3
5. t-Butyl Perbenzoate	1
6. MgO Dispersion	6
7. Glass, 1 in. Lengths	60%

B. Mix Properties

1. Masterbatch Viscosity   1312 cps

C. Physical Properties

1. Flexural Strength, 10^3 psi   50
2. Flexural Modulus, 10^6 psi   2.02
3. Tensile Strength, 10^3 psi   37.7
4. Tensile Modulus, 10^6 psi   2.41
5. Izod Impact, Notched, ft. lb./in.   21.7
6. Izod Impact, Unnotched, ft. lb./in.   30.3
7. Shrinkage, mil/in.   0

Low Pressure Compounds

Low pressure sheet molding compounds are tack-free, fully matured compounds that are moldable at plateau viscosities of 10–25 million centipoises rather than the 50–80 million centipoises normally encountered. The low viscosity condition can be obtained on some systems merely by reducing the quantity of Group II-A chemical thickener in the compound. However, the control parameters required for such a change frequently are impractical for production operations.

Forsythe and Ampthor[301] first described a controlled viscosity system using Paraplex™ CM-201 with the Paraplex™ P-340 system and a highly reactive magnesium oxide thickener. This paper states that the desired viscosity levels can be selected and controlled over a wide range, for example, 10–80 million centipoises. Basically the amount of MgO used determines the ultimate viscosity obtained, while the amount of maturation control agent regulates the speed of thickening and how quickly and how well the viscosity levels off. United States patent 3,879,318 was issued on that control method. The data sheet on Paraplex™ CM-210 states that the major benefits obtained by using the material in a formulation are as follows:

1. Reduced time between SMC preparation and the achievement of a moldable viscosity.
2. The ability to obtain a selected viscosity and to maintain that viscosity for extended periods.
3. More uniform molding characteristics of the SMC as a function of aging time.

It is significant to note that nowhere is molding pressure mentioned. The recommended level of maturation control agent and chemical thickener mentioned in those documents takes the compound to the normal viscosities one is assustomed to handling. The data clearly show and subsequent work has demonstrated that a lower molding viscosity window is easily obtainable by the proper selection of maturation control agent and MgO, and one would expect that a lower molding pressure would be required, but that fact was apparently ignored.

In 1971, Alvey[233] made a study of the effects of water on the MgO reaction in polyester resins and arrived at the following conclusions:

1. Excess amounts of MgO were necessary for chemical thickening.
2. Water increases the initial rate of thickening but reduces the overall level obtained.

The first resin vendor to actually promote low pressure sheet molding compounds was PPG Industries with its LMC™. That material was recommended for large-area automotive parts that could be molded at a reduced pressure.

Williams[371] announced Owens-Corning Fiberglas Corporation's low-pressure SMC product, SMC-II, at the 1976 SPI RP/CI meeting in Washington. The system was used on several truck front ends and on some commercial applications. The system did not have a class A surface.

High Strength Compounds

Structural grade sheet molding compounds generally are made by eliminating or drastically reducing the thermoplastic additive, using little or no inert fillers, and increasing the glass reinforcement content to the 50–70% range.

Rapp reported on work completed at the Amoco Chemicals Corp. Technical Service Laboratory on high strength molding compounds using various isophthalic polyester resins. That work showed that hard glass rovings permitted excessive resin flow-out during sheet preparation and during molding allowing the resin to flow away from the glass. On the other hand, soft glass rovings of the type used extensively in pigmented formulations showed a much greater resin holding capacity and excellent flow during molding. Filamentizing the rovings by passing them through several eyelets also increased the resin holding capability. Powdered MgO was initially used in this study, but it produced erratic thickening responses. When the thickener was changed to a dispersed MgO, thickening was reproducible. Formulation SMC-13 was used in this study, and its formulation and physical properties will be found elsewhere in this chapter.

Although the SMC-13 formulation, without inert filler, was proven to be a useful structural grade compound, it had some disadvantages, such as brittleness, a marginal dimensional stability, and a relatively poor surface finish. Including calcium carbonate filler in loadings of from 10 to 100 parts filler per 100 parts resin was then investigated. It was found that it was possible to incorporate 40 parts calcium carbonate filler and 65% glass loading and still retain good wetting of the glass. Flexural and tensile strengths were maximized at a filler concentration of 20 parts per 100 parts resin. However, the greatest overall improvement in impact strength, moduli, and shrinkage control came at a concentration of 40 parts filler. Formulation SMC-14 contains the added filler content, together with the physical properties of this formulation.

An attempt was also made to improve surface finish by the addition of thermoplastic additives. It was possible to prepare a zero shrink formulation by including 15 parts LP-40A to 85 parts polyester resin and 20 parts calcium carbonate. This formulation appears as SMC-15.

Low Density Molding Compounds

There is a great interest in reducing the weight of BMC and SMC molded parts, particularly in automotive applications. Glass microballoons are the principle material being used to effect that weight reduction, although other materials are also being considered. Chief problems with the use of glass microballoons are their relatively high cost and the loss of physical properties invariably associated with their use.

When glass microballoons are incorporated into a formulation, the inert filler loading generally is reduced to make resin available for coating the new filler.

Most FRP vendor laboratories have explored using glass microballoons in combination with their products and have published technical papers and reports covering this work. One such report concludes that an 8% loading of 3M B-40BX glass balloons in a Paraplex™ P-19D SMC results in a 20% reduction in compound density, satisfactory release and flow properties, very good surface smoothness properties, and adequate physical properties. The formulations from which those conclusions were made are shown elsewhere as SMC-16 and SMC-17.

Thompson, Weinman and Company made a BMC study using glass microballoons, keeping the filler volume constant. Some of their work is shown here as BMC-9, BMC-10, and BMC-11. The specific gravity dropped from 1.95 for the control batch (no glass microballoons) to 1.75 when the calcium carbonate was reduced from 220 parts to 165 parts per 100 parts resin and 7.5 parts glass microballoons were included or a 10% reduction. When an additional 55 parts calcium carbonate was removed and the glass balloon loading was doubled to 15 parts per 100 parts resin, the specific gravity dropped to 1.53 or another reduction of 12.5%. There also was a steady drop in physical properties as the glass microballoons were included. In an attempt to recover the lost physical properties, another batch of compounds was prepared, but the calcium carbonate was changed to a smaller particle size. Those formulations are shown as BMC-12 through BMC-14. The initial batch, BMC-12, is the control batch and does not contain any glass microballoons. Batch BMC-13 has 55 parts calcium carbonate removed and 7.5 parts glass microballoons included. The specific gravity dropped from 1.97 to 1.78, or slightly more than 10%. With that change the physical properties did not fall off as in the previous series but actually gained very slightly. In batch BMC-14 the filler concentration was maintained at the same level as in batch BMC-13 and the glass microballoon concentration was increased to 10 parts per 100 parts resin. The specific gravity dropped to 1.70 or another 4.5%.

CHAPTER 6

Molding Compound Process Equipment

SMC Impregnators

Three basic types of SMC impregnators are in general use today: nip rolls used with glass mat reinforcement; belt type used with roving and glass mat reinforcement; and the hollow can or roll type used primarily with roving reinforcement.

Glass Mat Reinforced SMC Impregnators

The first SMC impregnator design to appear in the literature was included in a paper by Scholtis.[77] Figure 6.1 shows a schematic line drawing of that impregnator. Glass mat is pulled through an impregnating tank after which excess resin is mechanically squeezed out and the wet mat is deposited onto a moving belt surrounded by a heated chamber. Film is applied only to the bottom side of the product after it exits from the heated chamber and as it is being wound onto the take-up roll. The problems encountered with the unit are obvious.

Gruenwald and Walker[337] discussed the evolution in West Germany of what they called "first generation" SMC impregnators, and those units are shown diagrammatically in Figure 6.2. The initial machine, Type I, consisted of a pair of driven rolls with a narrow gap between them. The carrier film and the reinforcing mats are fed into the nip between the rolls into which resin mix also is pumped. The rolls force the impregnation of the mat while sandwiched between polyethylene films

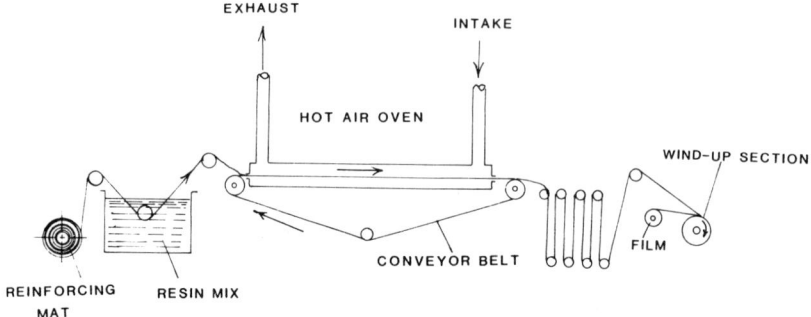

Figure 6.1 Schematic Diagram of the First SMC Impregnator in the Literature

and remove some of the entrapped air. Dams on each end of the rolls prevent resin mix from running out. That is a much-improved impregnator as compared with that mentioned earlier. Several plies of mat can be fed into the unit at one time to make a heavier product. There was a tendency for loose glass fibers to accumulate in the resin mix. Those glass fiber agglomerates interfered with impregnation, and when they were drawn into the roll nip, they tore the glass mat and some times damaged the rolls. These problems were overcome by installing a wide slit hopper above the rolls to introduce the resin mix between the rolls, as shown in Figure 6.2-II. Of course that greatly restricted the amount of resin mix that could be fed to the machine. As requirements for sheet molding compounds (SMC) increased, it became necessary to speed up the material travel through the impregnator, and unwet glass areas resulted. Increasing the roll diameters had no appreciable effect. Doctor blades then were installed over idler rolls to coat the polyethylene films with a layer of resin mix, as shown in Figure 6.2-III. While that change solved the dry glass problem, slippage then was encountered between the main driven rolls and the polyethylene sandwich. The problem was solved by texturizing or roughening the rolls.

Another type of impregnator designed to use reinforcing mats was built for the Dow Chemical Company's Freeport laboratory. A schematic diagram of that unit is shown in Figure 6.3.

Molding Compound Process Equipment 149

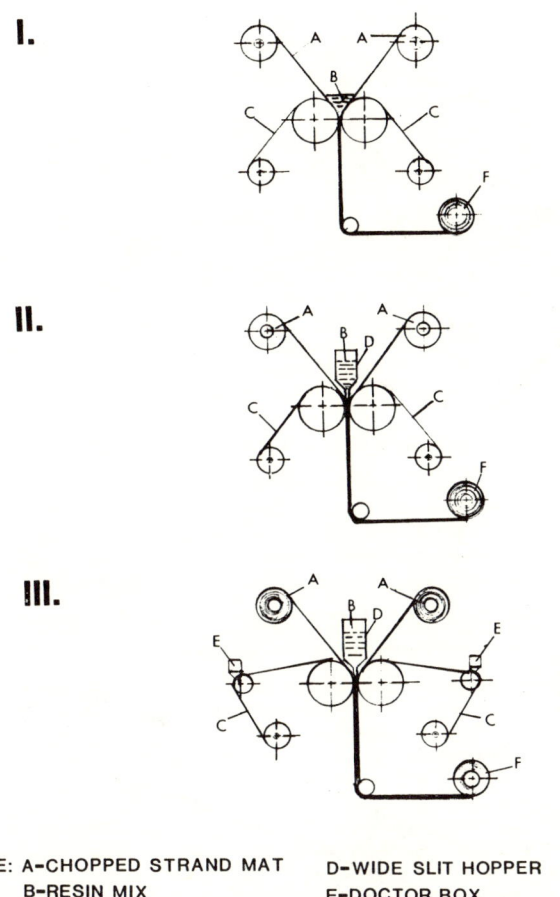

CODE: A—CHOPPED STRAND MAT D—WIDE SLIT HOPPER
B—RESIN MIX E—DOCTOR BOX
C—FILM F—SMC

Figure 6.2 Evolution of First-Generation Impregnators

The nip rolls were replaced with a pair of sizing belts, and the resin mix was doctored individually onto the two carrier films. Those films then were combined with the reinforcing mats between the sizing belts, which applied pressure to wet out the mats.

Belt Type SMC Impregnators

Glass reinforcing mats contain 5–20% binder to hold the glass fibers together. It has long been realized that the binder is an

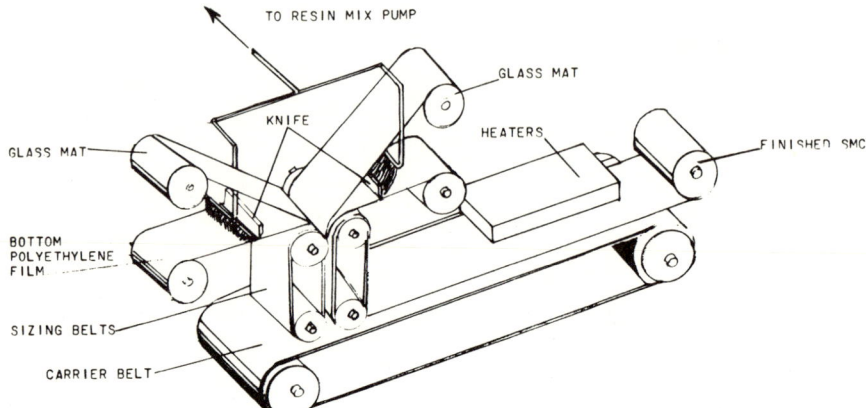

Figure 6.3 SMC Impregnator; Dow Design Using Mat

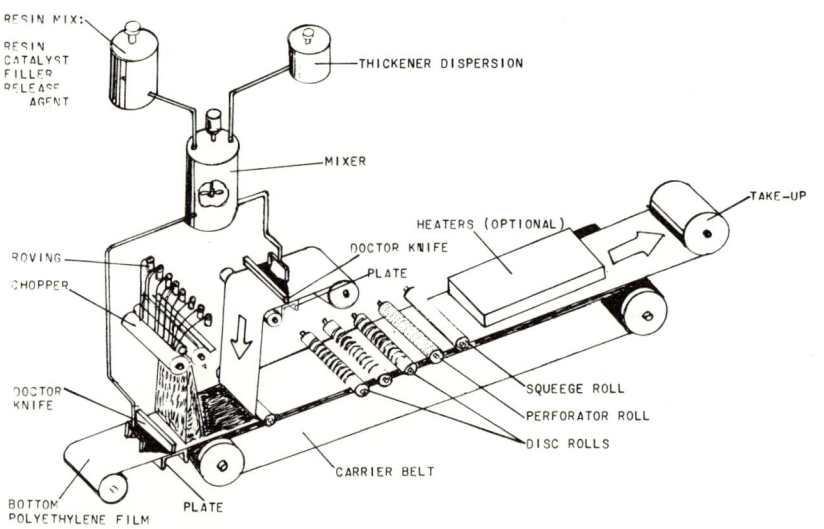

Figure 6.4 Belt Type SMC Impregnator

unnecessary part of the SMC product and actually may hinder proper wetting of the glass fibers in order to achieve maximum physical properties. Also, there was a limited quantity of reinforcing mats available when SMC development programs began, and none has since been developed. Conversely, there were numerous types of glass rovings available, and most fiberglass manufacturers have been very active in the development of special rovings to meet the requirements of SMC.

Several companies worked on the problem of incorporating roving directly into SMC. Wide roving cutters were available in the United States as early as 1951 from the I. G. Brenner Company. Gruenwald and Walker[337] mentioned that wide roving cutters were employed on the Bayer SMC impregnator in 1965 with the help of U.S. Rubber personnel. Owens-Corning Fiberglas Corporation is generally credited with building the first laboratory model in the United States that successfully used roving. Figure 6.4 contains a sketch of a typical belt-type SMC impregnator. Following the completion of the moving sandwich (after the top film has been applied) working disk rolls are included to force the resin mix into the glass fibers. Some machine designs use widely spaced disks at first and follow those with more closely spaced disks. A perforating roll was included on early machines to pierce the top carrier film to allow air to escape. In practice it was found that the paste quickly seals the perforations and does not allow much air to escape. Most perforating rolls have since been removed from impregnators.

Roll Type SMC Impregnators

There were some problems with belt type machines. First, it was necessary to keep the belt tracking properly, and while that technology is old and readily available, it does add considerably to the cost of the machine. Second, the SMC product was worked only on the top side. It was felt that working the SMC sandwich top and bottom would assure better wet-out. PPG Industries developed a can or roll-type SMC impregnator that works very well; a patent was issued on this unit to Potkanowicz[218]. Since the rolls are hollow, it is simple to fit them with rotary joints and to pump heat transfer fluid through the rolls to control their temperature. Some machine designs call for the rolls to be of slightly smaller diameters as the product

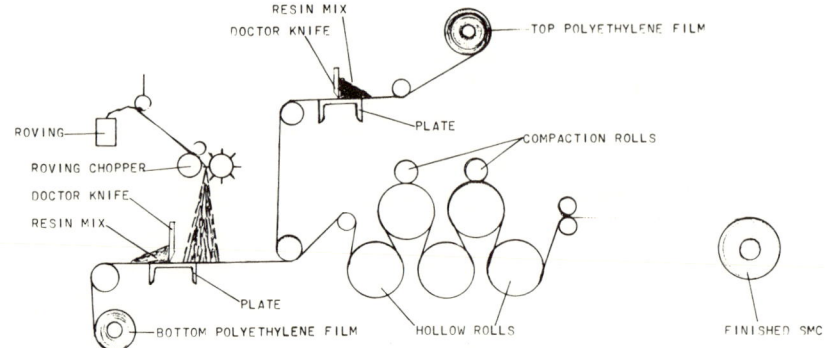

Figure 6.5   Roll Type SMC Impregnator

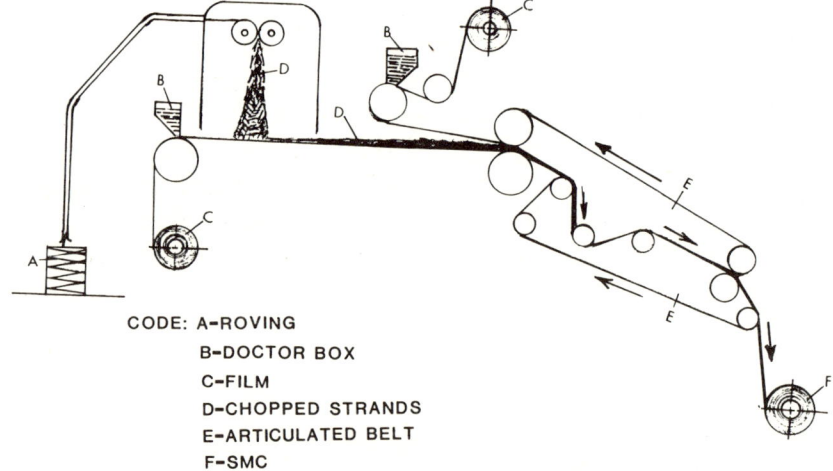

CODE: A-ROVING
B-DOCTOR BOX
C-FILM
D-CHOPPED STRANDS
E-ARTICULATED BELT
F-SMC

Figure 6.6   West German Articulated Belt-Type Impregnator

proceeds downstream so as to lightly stretch the carrier films. It also is simple to place working rolls so that they contact both sides of the sandwich. Figure 6.5 contains a sketch of one type of roll SMC impregnator.

Gruenwald and Walker[337] discussed the development of "second generation" SMC impregnators in West Germany in which chopped rovings were used as the glass fiber reinforcement with a variety of machine designs. In one such design the carrier belt was removed, and hollow rolls were added over which

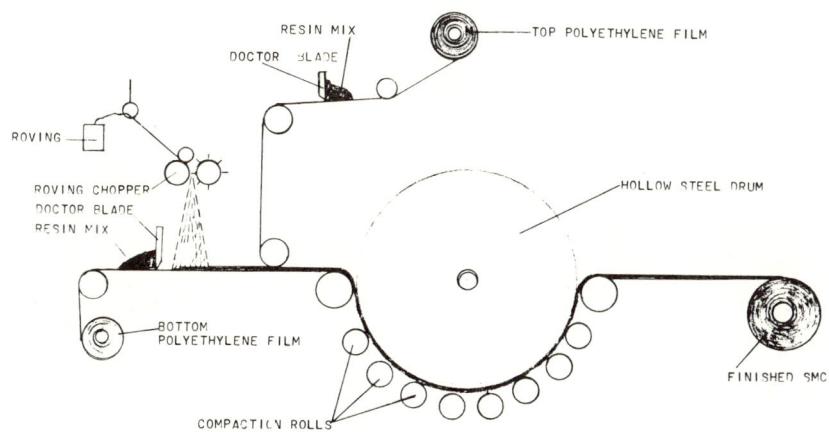

Figure 6.7 Hollow-Drum-Type SMC Impregnator

articulated belts worked the SMC product. A sketch of such a unit is shown as Figure 6.6. As far as is known, no similar machine was built in the United States and it appears not to be a popular type design. Another special type roll machine was discussed by Gruenwald and Walker.[337] It consisted of one large hollow drum over which the SMC sandwich was drawn as it is worked by a series of pressure rollers. One such type design is shown in Figure 6.7. One of those drum-type SMC impregnators was built and is still operating in the United States. One of the attractive features of roll-type machines is the fact that they require very little floor space. Most SMC manufacturers in the United States have available both belt-type and roll-type designs to satisfy the varied requirements of their customers. A number of optional features also are available on newer impregnators. Figure 6.8 consists of a sketch of a current model E. B. Blue Company roll-type SMC impregnator with several optional features. The top part of the sketch consists of a line drawing of the unit, showing the flow through the machine. The bottom portion is an elevation sketch of the machine. Three optional features are included: a self-contained resin-dispensing system, a stack of 9 heated rolls, and the wind-up section, being either a standard turret rewind or the newer festooning unit.

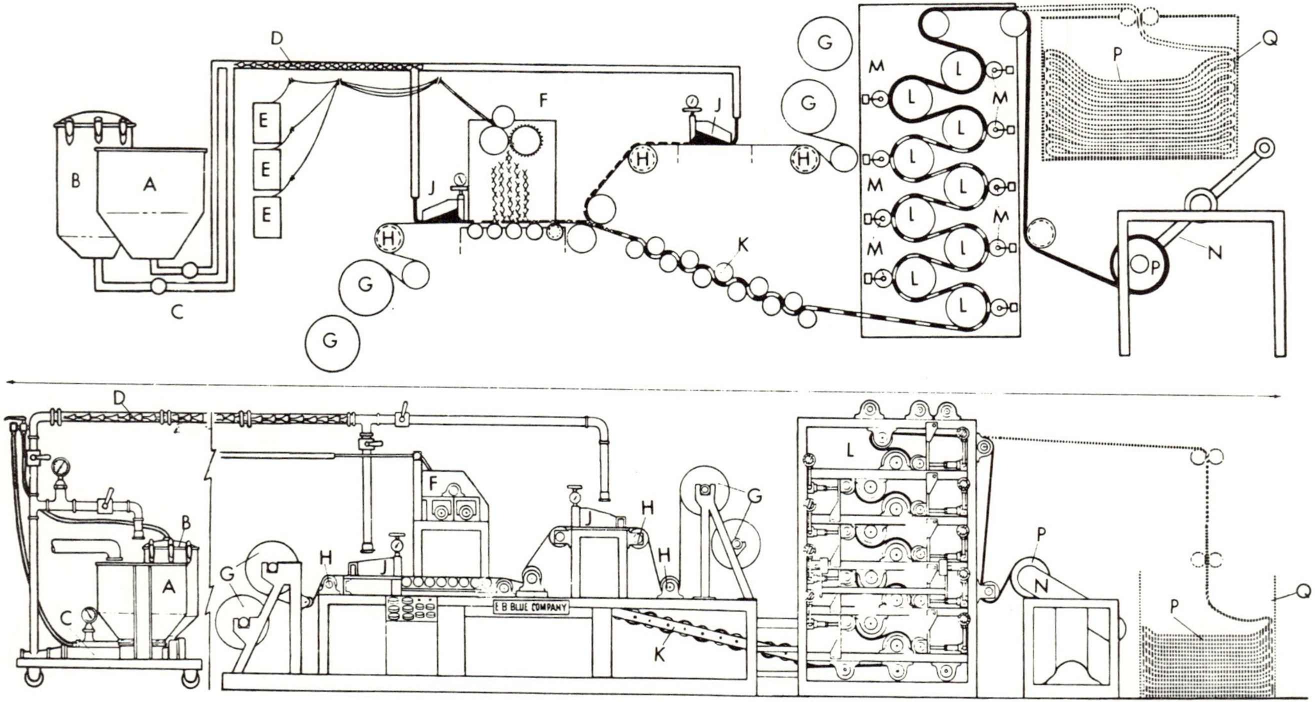

154   Handbook of Polyester Molding Compounds and Technology

A—TANK 'A' SIDE
B—TANK FOR THICKENER SIDE
C—POSITIVE DISPLACEMENT PUMP
D—STATIC MIXER
E—ROVING BALLS

F—ROVING CUTTER
G—FILM CARRIER
H—EXPANDER ROLL
J—DOCTOR BOXES
K—FIRST WORKING STATION

L—MAIN WORKING ROLLS
M—PRESSURE ROLLS
N—REWIND STAND
P—SMC PRODUCT
Q—FESTOONING UNIT

Figure 6.8 E. B. Blue SMC Impregnator

In this model the pumping system consists of two stainless steel holding tanks each with a Moyno™ positive displacement pump moving material through stainless steel sanitary quick-disconnect fittings. The two streams are combined and are run through a stainless steel static mixer, after which they are split and run to the two doctor boxes. Both pumps are equipped with variable speed drives. The stack of hollow rolls is variable with 9 rolls being standard. The SMC sandwich leaves the impregnator and enters the roll stack on the bottom, progressing vertically upward from one roll to the next. With 9 rolls the SMC is in contact with approximately 200 linear inches of heated roll surface. Each of the 9 rolls contains a small pressure-loaded work roll. Therefore, four work rolls contact one surface of the SMC product, and 5 contact the other surface. All the rolls are driven and are synchronized with the speed of the other impregnator units.

From the top of the stack of heated rolls the SMC product goes to the wind-up section, which may consist of the conventional 2-station turret rewind or the newer festooning unit. The first festooning unit was developed at the Granville laboratory of Owens-Corning Fiberglas Corporation to ship SMC to its Huntsville plant when the roof of the impregnating unit was blown off in a tornado. Using the festooning unit permitted some savings in floor space and shipping space and in the elimination of the wrapping film. In practice some rejected SMC product resulted since the folded ends tended to climb up the sides of the box, and to prevent that, severe tamping was done by the operators. That caused some boxes to have split SMC product, which was difficult to use. The inside of the boxes had to be coated to prevent the SMC from sticking to them. Several manufacturers now offer festooning units as optional units. Another optional unit is an in-line slitter, which is mounted between the turret wind-up and the impregnator. With that equipment the side trim scrap is cut off and discarded before the SMC product is rolled. Slits across the width of the SMC sandwich can be positioned as desired. That saves time later when the charge is prepared by the press operator.

SMC Impregnator Design Considerations

Framework

The framework should be sturdy enough to support all auxiliary units without buckling or deflecting. Welded structural angle and channel iron members generally are used for production impregnators, where a minimum of machine changes would be expected. The side member weldments are bolted together so that the unit may be disassembled for movement to another location or for maintenance when necessary. Provisions should be made to electrically connect all frame sections before painting to afford a good electrical ground to help dissipate static charges. Where space permits, extra framework should be provided for the installation of future accessories. For laboratory impregnators using metal framing channel systems, Unistrut™ or the equivalent is ideal, since changes can be made quickly and at little expense as the need arises. When those channel sections are used, the top channels should be fitted with removable closure strips to prevent dirt, chopped glass fibers, resin mix, etc., from getting into the open channel members. All frame members should be provided with leveling means to overcome the uneven floors frequently encountered in manufacturing areas.

Roving Choppers

The supports for the roving chopper should be mounted directly onto the main impregnator framework. The cutting width of the chopper should be 3 inches wider than the maximum SMC width to be produced. It has been found that for 1–2 inch cut fiber lengths the optimum height from the chopper roll centerline to the bottom polyethylene film is 24 inches. In many early SMC impregnators, especially laboratory models, provisions were included for adjusting the cutter height over the belt or film. That practice no longer is followed. The chopper must be provided with a variable speed drive and, at least for experimental units, should be independent of the main drive.

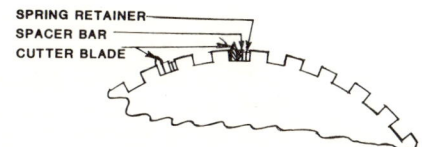

Figure 6.9 Finn & Fram Method of Holding Blades in a Cutter

On some production machines the chopper drive and main drive are coupled together so that both change together when an adjustment is made. An electrical brake should be provided to stop the cutter immediately when power is turned off.

On early roving choppers the cutting blades were ⅛-inch thick, high-speed tool steel that frequently required stoning to keep them sharp. Most modern roving choppers use Schick™ injector razor blade stock for the cutting edges. They are inexpensive and can be readily replaced. There are two general methods for holding those blades in place on the cutter roll. Finn & Fram uses an aluminum rectangular cross sectional spacer bar with a corrugated spring steel retainer in a rectangular slot machined into the steel cutter roll surface, as shown in Figure 6.9. The I. G. Brenner Company uses an aluminum wedge, as shown in Figure 6.10, to retain the cutter blades. The diameter of the metal cutter roll was chosen so that when its circumference is divided into 28 equal sections, half of which are milled slots, the width of the blade circumference is approximately ½ inch. Therefore, if two blades are put into each slot, the chopper will cut ½-inch lengths. If only one blade is put into each slot, the chopper will cut 1-inch lengths. Various combinations that add up to 28 ½-inch lengths can be cut at each revolution of the chopper roll. Both chopper designs work satisfactorily, but it is easier for inexperienced personnel to change blades in the Brenner design than in the Finn & Fram design. Both cutters include moving combs, which continually change the point where roving enters the chopper to prolong blade and rubber roll life. Both designs make use of replaceable rubber cots on the rubber back up rolls. It has been found to be profitable on long rubber rolls to provide a spare rubber roll mandrel that is knurled and to have rubber stock wound onto

158 Handbook of Polyester Molding Compounds and Technology

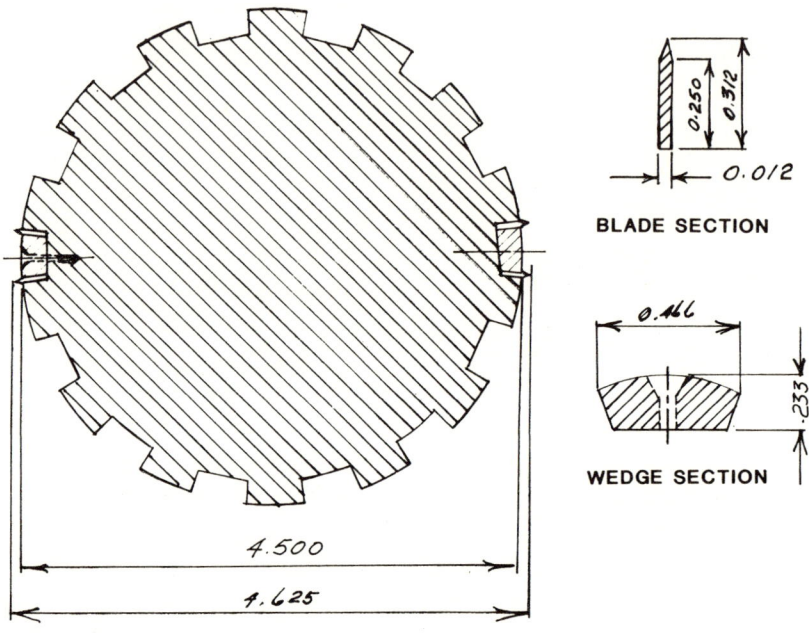

Figure 6.10 Brenner Cutter Roll Design

this mandrel, cured, and precision ground to the proper diameter. That procedure will extend the life of the rubber roll considerably as well as reduce maintenance costs.

Polyethylene Film Stations

Two polyethylene film let-off stations hold the supply of film and feed it at controlled tension to the two doctor blade areas. For production impregnators each station should be able to hold two rolls of film to minimize downtime when 1 roll is exhausted. A 19½ inch diameter roll of 2 mil polyethylene film will last for approximately 10 hours continuous production time at a speed of 20 feet per minute.[212] Proper film tension is provided by adjustable pneumatic or mechanical brakes working on the film roll mandrels. No breaking force should be applied directly to the polyethylene film itself because it would distort and possibly tear the film. Expander rolls immediately in front of the doctor blades will keep the film flat and free of wrinkles. Small rubber-covered toe-out rolls, placed at a slight angle to the film, travel at its outer edges near the doctor boxes, also have been useful but are not so successful as expander rolls.

Doctor Blade Stations

At two locations, a doctor blade meters the resin mix onto the carrier film as it is drawn between the blade and its base plate. The doctor blade itself should be steel and as thin as practical without distorting. The doctor blade should be adjustable in a vertical plane. The blade should be ground on the reverse or downstream side to a dull knife-blade edge. Radiused blades do not provide a linear output as the gap is opened. All attachments should be made of aluminum for weight reduction since they must be manually removed and cleaned. Simple adjustable side dams are satisfactory for small laboratory machines. An enclosed box design is preferred for production machines with the doctor blade forming one side of the box.

One doctor box design uses an inclined plane with micrometer adjustments at one end of the box. Another design has a hinged base plate, which permits the plate to be raised when lumps in the resin mix threaten to tear the film.

Combining Section

Whether the impregnator be of the belt type or the hollow roll type the resin mix coated top carrier film must be combined with the bottom coated film and the cut reinforcement fibers. The angle at which that top film approaches the bottom film is critical; too shallow an inclined angle disturbs the glass; a too open angle draws more air into the sandwich. It has been found that the optimum angle is approximately 20°, as shown in Figure 6.11.

Drive

A variable speed drive is required for both belt type and hollow roll-type impregnators. For production machines a DC drive is best if one can afford the expense since the speed variations

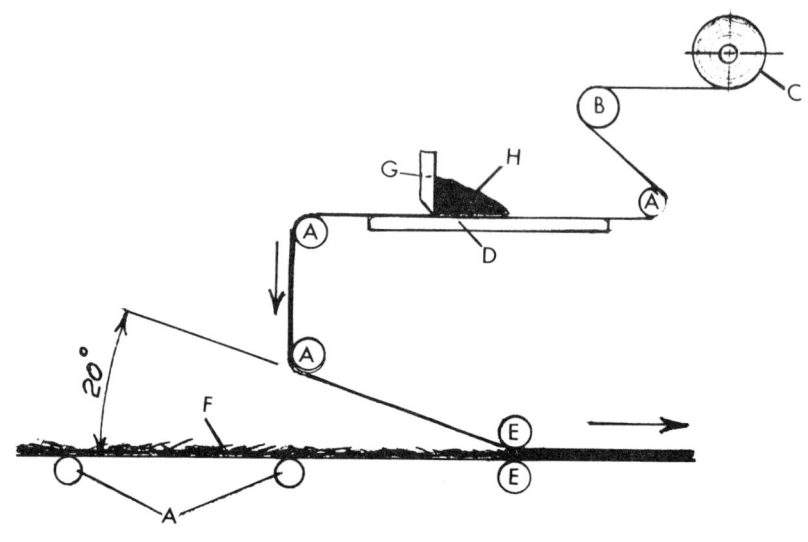

A—IDLER ROLL
B—EXPANDER ROLL
C—FILM
D—BASE PLATE

E—COMBINING ROLLS
F—UNCOVERED SMC
G—DOCTOR BLADE
H—RESIN MIX

Figure 6.11 Schematic Diagram of Combining Section

are endless and are easily changed and controlled. Most impregnators use a mechanical variable speed control, such as a Reeves Motodrive with a speed ratio of at least 4:1. Setting the lower speed at 7.5 feet per minute (fpm) would provide a top speed of 30 fpm, and that nicely brackets the 20–25 fpm range most often used in production.

Rolls Or Conveyor Belt

If a belt-type impregnator is to be considered, the conveyor belt should contain a top surface that is impervious to styrene. A Teflon™ coated 2-ply fabric belt is best, and it should be purchased as a bias spliced endless belt at least four inches wider than the maximum width sheet to be produced. Neoprene coated fabric belts also are in use, but care must be exercised in cleaning severe spills since the rubber will absorb styrene and solvents and expand. The belt must be adequately supported with idler rolls, and a tracking control system must be provided. Crowned drive rolls should not be used.

On hollow rolls that are heated, aluminum rolls 10 inches in diameter have been found to be satisfactory and have excellent heat transfer. Steel rolls may be used, but additional heating surface should be provided. Those rolls must be driven and pneumatically loaded. Teflon™-coated steel work rolls equipped with adjustable stops are preferred. In the event of a film tear, the resin mix quickly thickens on the heated rolls and is easily peeled off. Teflon™ coating the steel rools materially simplifies clean up and shortens downtime. The work rolls are serrated, and generally wide serrations are used on the first roll followed by more closely spaced serrations as the SMC proceeds downstream.

Take-Up Section

The take-up section winds the finished SMC product onto tubes under controlled tension. It may consist of one stationary position, two stationary positions, or a two-position turret winder. With single position winders the impregnator must be stopped at the completion of each roll. While that is satisfactory for laboratory experimental runs, for production operations either the two-position stationary winder or the turret winder are required; both of those units allow continuous operation. A means must be provided to wrap the completed SMC roll and to move it to a storage rack or skid. A movable overhead crane mounted on an I-beam generally is used. Handy accessories include a footage counter or a scale on the overhead crane or both to help keep the size of finished rolls uniform. The footage counter also can be used with a watch or clock to accurately check the impregnator line speed.

162 Handbook of Polyester Molding Compounds and Technology

Figure 6.12 Roll Type SMC Impregnator With Turret Wind-Up and In-Line Slitter (*Courtesy: E. B. Blue Co.*)

Commercial SMC Impregnators

E. B. Blue Company

The E. B. Blue Company has been making SMC impregnators as long as anyone else in the United States. Figure 6.8 is a schematic diagram of one of its more advanced SMC machines. Figure 6.12 is an actual photograph of one such machine installation. In the foreground of this photograph is the turret wind-up station with a slitting station upstream. The stack of heated rolls and work rolls are further upstream followed by the top film station, single roving cutter, bottom film station, and, in the background, the roving reinforcement supply station. The movable overhead crane is used to install new film rolls and for routine cutter maintenance.

Figure 6.13 shows a much narrower E. B. Blue SMC machine being checked out at the factory before shipment to a customer.

Figure 6.13 Narrow Laboratory Type SMC Impregnator With Adjustable Doctor Box (*Courtesy: E. B. Blue Co.*)

The view is taken from the downstream end of the machine, looking upstream, and clearly shows the adjustable doctor box station with a slat expander immediately in front of it to smooth out the carrier film. An operator passageway, which permits easy access to both sides of the machine, is in the foreground.

Figure 6.14 is a close-up view of the drive section of a pair of belt wet-out roller stations that work the SMC product for better reinforcement wet-out and higher physical properties in the molded part. That section is particularly useful on high glass content compounds and is an option that can be added to an SMC machine.

I. G. Brenner Company

Figure 6.15 is an overall view of an I. G. Brenner Company roll-type SMC machine. In the left foreground a hot water heater is shown for running the hollow rolls at an elevated temperature

164   Handbook of Polyester Molding Compounds and Technology

Figure 6.14   Close-up of Drive Section of Belt Comparison Unit (*Courtesy: E. B. Blue Co.*)

to improve glass wet-out and hasten the SMC maturation reaction. Belt type SMC machines are also available, as are several optional pieces of equipment.

Engineering Technology Inc.

Figure 6.16 consists of a view looking downstream on a 60-inch wide Engineering Technology SMC machine. The rear film station consists of dual sleds mounted on wheels that roll on tracks for easy film replacement and servicing. The adjustable doctor box station is surrounded with a hood to collect styrene fumes and to channel them into the framework, which consists of a large cabinet for collecting and disposing of such fumes. The cutting section consists of dual 60-inch wide roving choppers. A roll type expander for eliminating film creases is shown immediately above the rear film station.

Figure 6.15 Brenner Roll Type SMC Machine (*Courtesy: I. G. Brenner Co.*)

Figure 6.16 EnTec 60 Inch SMC Machine (*Courtesty: Engineering Technology, Inc.*)

BMC Process Equipment

Double Arm Mixers

Intensive double arm batch mixers for preparing premixes and BMC are made by the Day Mixing Co., by Baker-Perkins, and several other equipment manufacturers. A variety of arms or blades are available, and some of the styles are shown in Figure 6.17. For the chemical and bakery industries those double arm mixers are supplied with sigma blades and with a blade-to-shell clearance of from 0.010–0.050 inches. It has been found that sigma blades entrap resin, which will not mix in with the balance of the mix. Sigma blades are also very difficult to clean. A dispersion type or single-curve blade is much more acceptable. The dispersion blade has sharp edges while the single-curve blade (either 135° or 180° curve) has no sharp edges and is considerably easier on the glass reinforcement during mixing. The two blades rotate toward each other tangentially with blade speed ratios of 3:2. For fiberglass reinforced molding compounds a blade-to-shell clearance of ⅛ inch is used for the small laboratory mixer with a capacity of 2¼ gallons and from ¼- to ⅜-inch clearance on 50- and 100-gallon production mixers. Figure 6.18 shows a 200-gallon production double arm mixer containing sigma blades, and Figure 6.19 shows a similar capacity mixer having dispersion blades. Both mixers have the automatic dumping feature.

Clay-Type Extruders

In order to densify BMC charges and get them into a handleable form, they are generally run through some type of extruder. Most hydraulic type extruders are designed and built in-house. Clay-type extruders, similar to those used for making bricks, are used for extruding BMC. Those extruders have very large diameter screws, which are polished and chrome plated to make them easier to clean. Some have pneumatic cutters built in to cut the extruded logs to length. Figure 6.20 shows a late production model clay-type extruder having a pneumatic cutoff.

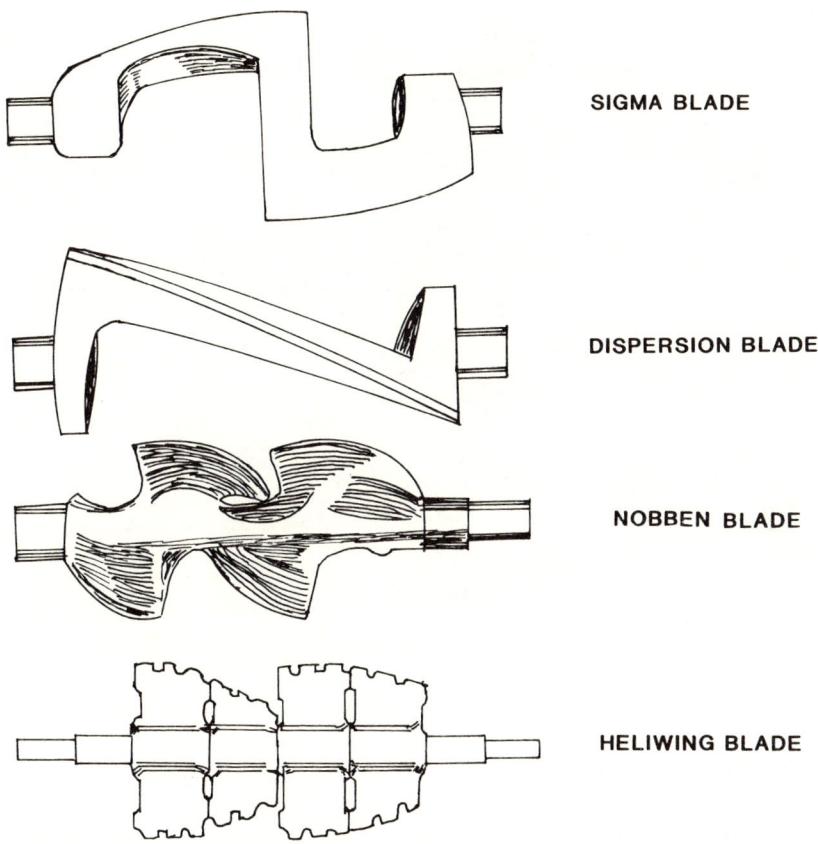

Figure 6.17 Several Double Arm Mixer Blade Styles

Table 6.1 contains data on several Bonnot clay-type extruders. Figure 6.21 shows an earlier model Bonnot clay-type extruder being used in production.

Miscellaneous

One unique piece of process equipment consists of the Ross/ AMK Kneader-Extruder (Charles Ross & Son Co.), as shown in Figure 6.22. This unit prepares the BMC in a conventional manner in a double arm mixer, but instead of dumping the

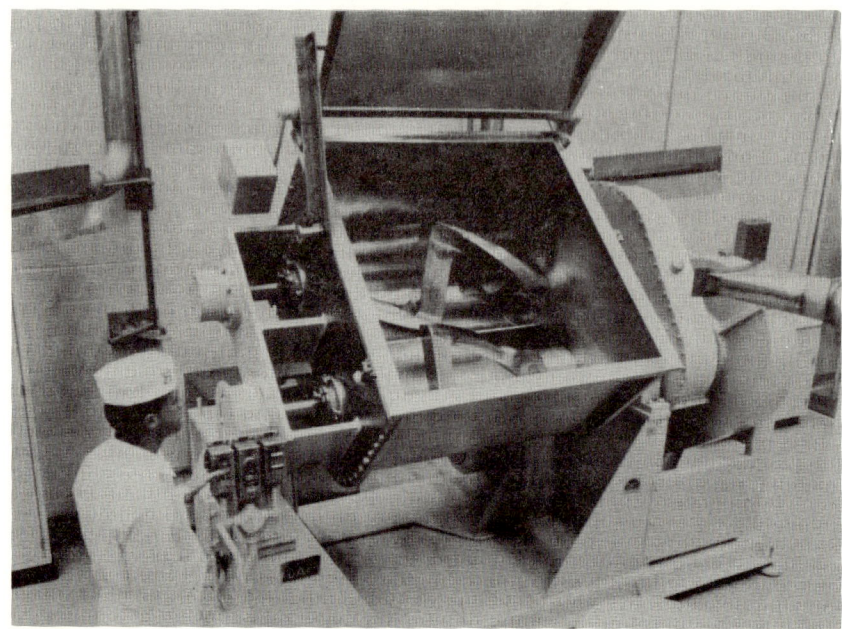

Figure 6.18 A 200-gallon Double Arm Mixer Having Sigma Blades (*Courtesy: Day Mixing Co.*)

completed mix for additional processing, the mix is extruded into log form from the mixer by an extrusion discharge screw located in the cavity beneath the mixing blades. A pneumatic guillotine cutter can be added to automatically cut the extruded logs to length. The unit combines the efficiency of the double arm mixer with the convenience of the extrusion screw. Some materials handling problems are therefore eliminated, and the risk of operator injury is minimized since the mix is not dumped. A variety of mixer sizes is available from 1–1,000 gallons working capacity.

The use of a Ross planetary mixer to prepare premix batches is another unusual application (such a mixer is shown in Figure 6.23). A decrease in total mixing time is reported, as is an improvement in compound strength owing to less fiber degradation. The low-viscosity ingredients are mixed at high speed, and the final mixing operation is done at a lower speed to

Figure 6.19 A 200-gallon Double Arm Mixer Having Dispersion Blades (*Courtesy: Day Mixing Co.*)

protect the glass reinforcement. Mixer sizes of from 1–300 gallons are available. No seals come into contact with any mix ingredient, so maintenance problems are minimal, as compared with other compound mixers. A vacuum can be supplied on some models to remove entrapped air from the mix.

One of the problems that has confronted the industry is the handling of premix and BMC materials. Densifying the charge by running it through a clay extruder is of great assistance for compression molding. Before BMC injection molding was a success, a means had to be found to transfer the bulk compound into the injection machine. Several different types of loading

170 Handbook of Polyester Molding Compounds and Technology

Figure 6.20 A Late Model Bonnot BMC Extruder (*Courtesy: The Bonnot Co.*)

equipment or stuffers were developed. One of the most satisfactory of the stuffers is made by Martin Hydraulics, and one model is shown in Figure 6.24. A hydraulic ram feeds the BMC into the injection molding machine. When the ram is retracted, it is loaded by means of a hopper mounted directly over an opening above the ram. Bulk molding compound inside the hopper is transferred to the ram opening by a close fitting plate that is attached to another hydraulic cylinder. Models are available to fit most injection molding machines.

Figure 6.25 shows an 1,850 ton down-acting SMC press made by John T. Hepburn, Ltd. of Toronto, Canada. Some of the specifications for this press are:

Press Capacity: 1,850 tons, adjustable
Overall width × depth × height: 15.5 × 8.5 × 30.5 ft.
Stroke: 0-48 in. adjustable
Platen Size: 120 × 92 ins. F to B
Fast Advance Speed: 50 ins./min.
Pressing Speed: 0-18 ins./min. adjustable
Breakaway Speed: 0-168 ins./min. adjustable
Fast Return Speed: 378 ins./min. maximum

Several options are available, including a mold transport system, an electro/hydraulic leveling and/or parallelism control system, and microcomputer control.

Figure 6.26 shows a typical SMC production press line.

TABLE 6.1 A Description of Several Bonnot BMC Extruders

Model Number	A	B	C	D
Nominal Capacity, Lbs/Hr	1,000	3,000	900	1,500
Worm Sizes, inches	8-6 or 4	10-8	6-4	8-6 or 4
Counterrotating feeder	no	yes	yes	yes
Motor Size, Hp	5	15	5	7½
Hopper Size, inches	8 × 29	11 ⅜ × 19	9 ¼ × 17	9 × 29
Extrusion Sizes, inches	1-4	1½-6	½-4	1-4
Common to All Models:	Variable Speed Drive Water-Jacketed Barrels Pneumatic Cutters Chrome-Plated Dies & Extrusion Worms Water Cooling of Extrusion Worm Optional on C & D			

Figure 6.21 An Early-Model Bonnot BMC Extruder in Production

Molding Compound Process Equipment 173

Figure 6.22 Ross/AMK Kneader-Extruder (*Courtesy: Charles Ross & San Co.*)

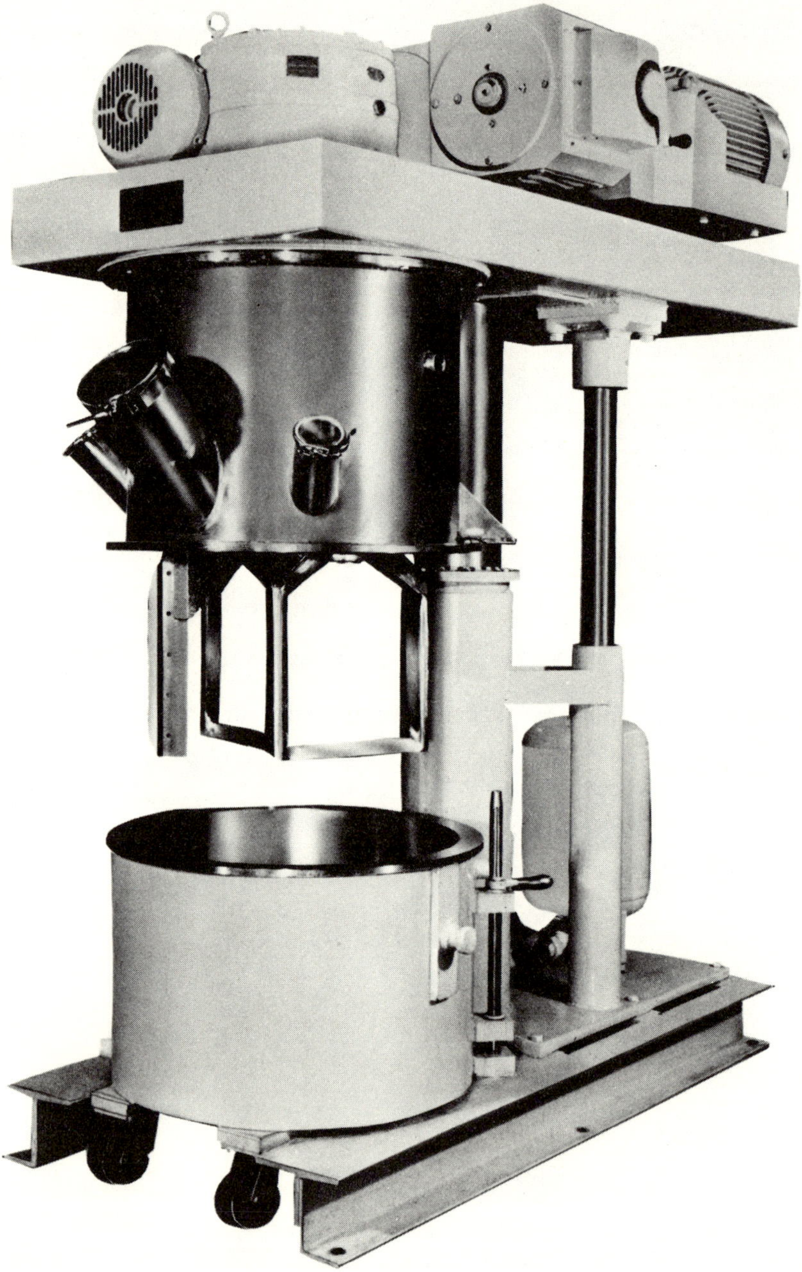

Figure 6.23 Ross Planetary Mixer (*Courtesy: Charles Ross & Son Co.*)

Figure 6.24 A Martin Hydraulics Stuffer Mounted on a Van Dorn Injection Molding Machine (*Courtesy: Martin Hydraulics*)

Figure 6.25 A Large Down Acting SMC Press (*Courtesy: John T. Hepburn, Ltd.*)

Figure 6.26 A Typical SMC Production Press Line (*Courtesy: Williams, White & Co.*)

CHAPTER 7

Compression Molding Technology

Introduction

It now is well understood that a satisfactory molded fiberglass reinforced plastic (FRP) part results only after the part has been adequately designed, a proper mold has been procured, and a suitable molding material is employed. Insufficient attention to any one of these three factors can result in a disaster for the molder. In the early days of compound development the primary emphasis was placed on the materials, since they were new and novel. Attempts to resolve problems generally took the form of materials changes. Some of the headaches incurred by molders were the result of their using poorly designed molds or trying to mold improperly designed parts.

Molding Compound Variables

Table 7.1 consists of a tabulation of molding compound variables. The first column contains a list of 54 independent variables all of which have some effect on the final molded part (there probably are other variables that have been overlooked). Some of the variables listed are critical in nature while other have only a very slight influence. From this list it should be seen that molding compounds are complicated materials. A thorough knowledge of the chemistry of the particular system

TABLE 7.1 Molding Compound Variables

I. Independent		II. In-process		III. Dependent
A.	*Materials*	A.	Resin Mix	1. Surface Smoothness
	1. Resin		Maturation	2. Surface Defects
	Molar Composition		Shelf Life	3. Fiber Orientation
	Water Content		Mix Viscosity	4. Repairability
	Free Glycol Content		Styrene Evaporation	5. Long-Term Waviness
	Reactivity		Mix Reactivity	6. Physical Property Uniformity
	Viscosity	B.	Temperature	7. Physical Properties
	Solids Content		SMC Conditions	8. Paintability
	Acid Number		Web speed	9. Bondability
	Hydroxyl Number		Air Entrapment	10. Economics
	Contaminants		Environment	
	2. Thermoplastic Additive		Glass Content	
	Type		Glass Length	
	Concentration	C.	Glass Distribution	
	Compatibility		Static Generation	
	3. Monomer		BMC Preparation	
	Type		Type Mixer	
	Concentration		Blade Style	
	4. Inhibitor		Glass Content	
	Type		Glass Distribution	
	Concentration	D.	Extruding/Compacting	
	5. Organic Peroxide		Compound Storage	
	Type		Temperature	
	Concentration		Humidity	

TABLE 7.1 Continued

I. Independent	II. In-process	III. Dependent
6. Inert Filler 　Type 　Concentration 　Particle Size 　Moisture Content 　Coating 　Contaminants 7. Colorants 　Type 　Concentration 　Effect on Reactivity 8. Release Agent 　Type 　Concentration 9. Thickening Agent 　Type 　Concentration 　Particle Size 　Reactivity 　Moisture Content 10. Glass Reinforcement 　Type 　Coupling Agent 　Binder Content 　Unit Weight 　Moisture Content	E. Molding 　Mold Temperature 　Cure Cycle	

TABLE 7.1 Continued

I. Independent	II. In-process	III. Dependent
B. Mechanical		
1. Part Design		
Geometry		
Projected Area		
Bosses		
Ribs		
2. Mold Design		
Shear Edge Height		
Shear Edge Clearance		
Surface Finish		
Effect of Stops		
3. Press Design		
Rate of Closure		
Parallelism		
Rigidity		
Tonnage		
C. Operational		
1. Charge Weight		
2. Charge Shape		
3. Charge Placement		

a molder plans to use is not absolutely necessary for successful molding, but that knowledge is very useful for troubleshooting plant molding problems. A knowledge of the materials and their interactions is necessary when revisions to an existing formulation are required in order to satisfy a new part specification.

The second column contains a listing of in-process variables that are greatly influenced by factors in the first column. For example, the resin mix viscosity is dependent on the type of polyester used, styrene content, filler loading, and in some cases by the type and concentration of release agent used. If a thickened system is used, the maturation characteristics of the compound are governed by the type of polyester used and the type and concentration of chemical thickening agent, while the rate of the maturation reaction is considerably influenced by the presence of free moisture in the resin and filler and free glycol in the resin.

The right-hand column of Table 7.1 contains a listing of the dependent variables or the responses resulting from the independent and in-process variables discussed earlier. They are listed in what is considered their order of importance. Surely the most important characteristic of low profile bulk molding compounds (BMC) or sheet molding compounds (SMC) is their superior surface finish. The lower-cost raw materials in BMC compared with SMC, preform, and mat moldings is an important consideration in their use in most applications. Physical properties are of less importance in BMC. Where physical properties are important SMC should be considered.

Surface defects in premix and mat or preform reinforced parts prevented the wider use of FRP for many years. The use of chemical thickening agents improved BMC parts and the inclusion of low profile additives in both SMC and BMC has resulted in further improvements. As those improvements occurred, the materials became stiffer and required more molding back pressure, but at the same time surface defects were reduced. If defects are unavoidable in a part, they must be capable of being repaired. A good example is a nick in the mold that results in

TABLE 7.2   More Important Molding Variables

1. Press Show Close
2. Molding Pressure
3. Mold Temperature
4. Cure Cycle
5. Charge Size
6. Charge Weight
7. Charge Shape
8. Charge Placement

a raised section on the part. Owing to the demand for parts, it may not be practical to stop production to repair the mold. A business decision may dictate that the part be continued in production and the defect be sanded out in a secondary operation.

Compression Molding Variables

Table 7.2 contains a list of the more important compression molding variables. Generally, they are factors that must be under control for a satisfactory molding operation. The first four items are dependent on the press setup and are generally outside the influence of the press operator. Controlling the last four items listed is generally the sole responsibility of the press operator. His attitude and the degree of reproducibility of those items directly affect the molded part defect rate.

Press slow close is considerably more important in molding BMC and SMC than it is for mat or preform molding. The objective is to flow the compound in a uniform front at a rate to allow the air in the cavity to escape through the shear edge without getting entrapped by the advancing fronts: too slow a closure can allow the material to pregel on the hot mold; too fast a closure will entrap air. Modern presses can close at a rate of 600 inches per minute to within one inch of the shear edge, at which point they are slowed down to a variable closure rate of 1–20 inches per minute. A press closure faster than 4 seconds for the final 1-inch travel might damage the shear edges. A slow close of 5–20 seconds for the final 1-inch travel is generally satisfactory. The time from the first compound contacting the hot mold to full pressure should not exceed 20 seconds.

Molding pressures of 100–300 psi on the parts projected area are usual for mat and preform parts. Molding pressures of 500–1,200 psi are required for the stiffer thickened compounds. A pressure of 1,000 psi on the projected part area is quite common. Higher pressures increase sink marks. Lower pressures tend to cause scumming of the mold and porosity.

Mold temperatures depend on the particular resin-catalyst system being used. In order to obtain a reasonable cure cycle, most materials are specified to be molded at approximately 300°F (149°C). Generally, it is advisable to maintain the female or cavity portion of the mold 10°F hotter than the male or core portion. Temperature variations within a hot mold portion should be maintained within ±10°F. To accomplish that, the mold must be properly designed for good heat transfer and steam traps should be mounted lower than the mold to prevent condensate buildup and must be kept free of scale. Other suggestions that amount to good practice in steam heating compression molds will be found in Table 7.3.

Minimum cure cycles are desirable. The point at which complete cure has been reached is a matter of conjecture. The use of bosses and ribs in BMC and SMC parts further complicates that conjecture. In breaking in a new mold, a safe cure time is customarily employed and is then progressively reduced. A rule-of-thumb of 1 second per mil of part thickness is a convenient guideline. Barcol hardness is used as a criteria of cure on mat and preform parts, but that method is of little value when low profile additives are used that automatically reduce the Barcol hardness. Some technicians favor quickly removing the part from the press area and checking it for styrene odor. The presence of styrene odor in a part is considered to be adequate proof of undercure. Overcure of mat or preform reinforced parts generally is considered to be only uneconomical. With BMC and SMC molded parts there is evidence that overcure contributes to parts sticking in the mold and to slight surface crazing.

TABLE 7.3 Fundamentals of Steam Heating Compression Molds

1. All presses should be piped individually off the main steam header.
2. Each mold section (plug and cavity) should be piped individually from the supply line with its own trap, controls, etc.
3. All steam supply lines should be insulated to minimize condensate forming in front of the mold being heated.
4. All piping (supply and return) should be of such size as to allow sufficient quantity of steam to pass through the mold for constant heating (a minimum pressure drop to the trap). All the piping between a temperature controller, through the mold, and to a trap should be in series only. Any parellel leg can become waterlogged and stop passing steam, thereby losing temperature at the mold area supplied by the leg.
5. The steam trap should be the lowest point in the heated system (lower than the mold).
6. All steam lines through the mold to the trap should be pitched to drain toward the trap. The flexible lines from the mold should not lie on the floor and then rise to the trap. There should be no low sections for condensate to collect.
7. Steam controllers and traps should have line strainers in front of them.
8. In high-pressure steam systems, the condensate return line may be located above the mold, and the condensate will be pushed up to it each time the trap opens. If 1 trap sticks open, it will pressurize the condensate return line to where other molds on the same system will not expel their water, losing mold temperature down the line.
9. Impulse traps require less maintenance than do bucket-type traps.
10. Strainers and traps should be inspected at least once a month to ensure that they are functioning. There are a number of ways to be assured that a trap is operating. It may be listened to with a screwdriver held against it, and the ear or an automative stethoscope. The trap opening and closing and condensate flow may be heard. A more precise method is thermocouple and temperature recorder attached to the outlet of the steam condensate line just downstream of the trap and covered with insulation. Every time the trap opens, the line temperature will raise. The chart of temperature versus time will show the functioning of the trap. There should be a sine curve formed by periodic condensate–steam flow. To further expound the use of the temperature recorder, 2 more thermocouples should be added, 1 at the main steam supply header and the other on the steam line inlet to the mold. The temperature at both locations should be the same. Should the temperature at the mold inlet be less than the header, the pressure drop is to a large extent due either to steam capacity or line size, too small between the header and the mold.
11. The mold temperature controllers should be receiving their input from the mold or right next to the exit from the mold, not several feet downstream. Many plants drill and tap both halves of the mold and insert thermocouple jacks.

Unfortunately, the selection of charge size, weight, shape, and placement is generally a matter of trial and error. Past experience counts considerably in being able to start a job in a minimum of time.

Determining the proper charge size is more complicated. Generally the BMC charge is compacted first by extruding to get it into a convenient log form. The larger the diameter of the log, the less loss of physical properties resulting from fiber degradation. On complicated and large parts sometimes two different size extrusions are used. A simple charge shape generally is best. Charge shape depends on the geometry of the part and can be adapted to fit the shape of the mold. When multiple extrusions are used, they usually should not be butted in the charge placement that might entrap air. Overlapping the extrusions generally is more satisfactory.

Small plugs of BMC should not be stuffed into bosses. That was a popular practice several years ago. Pyrolytic investigations of such sections on a molded part revealed the folded glass fibers with pockets that prevent the escape of entrapped air. It is preferred that the glass fibers flow into such cavities. When multiple charges are necessary, it is preferred that they be connected, if at all possible. In older presses, or on molds that deflect so that one corner closes before the others, it is generally necessary to place the charge on the corner that closes first so as to flow away from that area.

Testing Molding Compounds

Viscosity Increase of Thickened Resin Mixes

Initially, retained resin mix samples had their viscosity checked periodically, and a plot was prepared of viscosity versus time. The time-consuming nature of such checks has largely been eliminated by the use of the Brookfield recording Rheolog viscosimeter. Figure 7.1 contains such a viscosity recording for a fairly fast maturing system. Since the popular Brookfield viscosimeter has a maximum capacity of 8 million centipoises, it

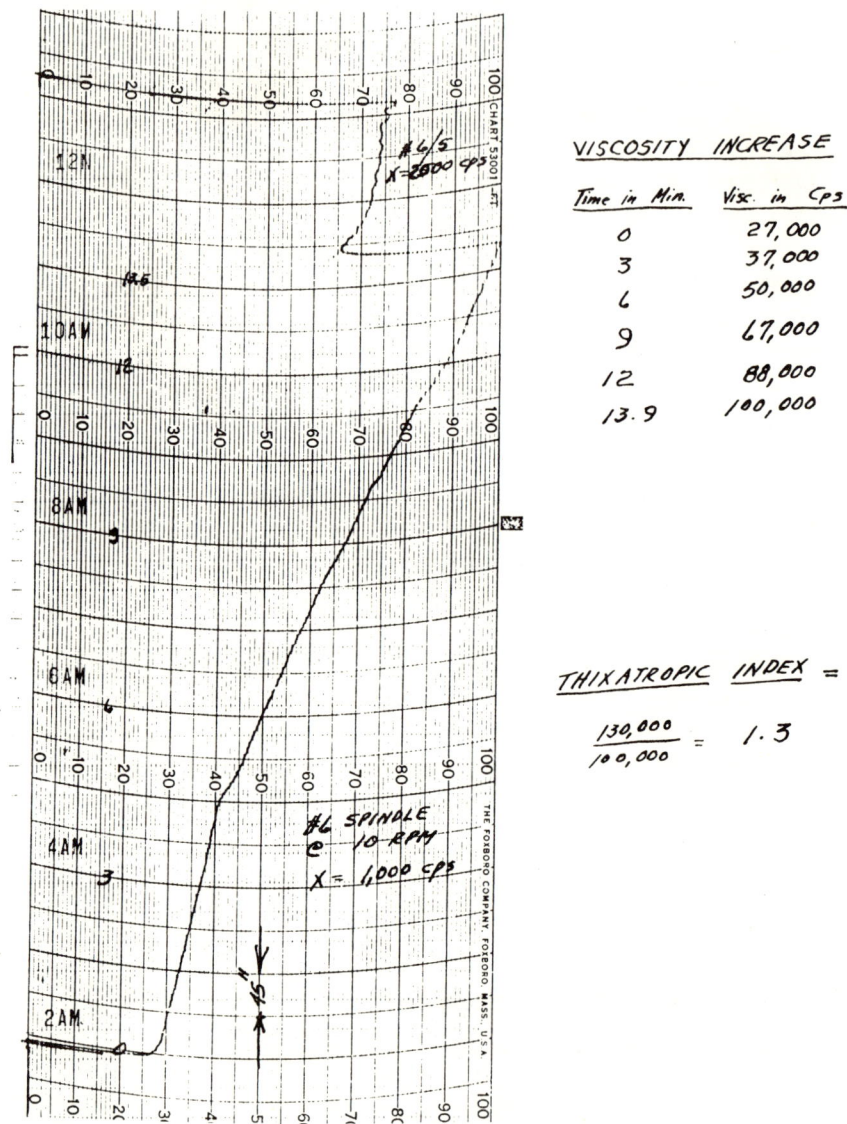

Figure 7.1 Continuous Plot of Viscosity of one SMC Mix

has been necessary to use T spindles on a model HBT Brookfield viscosimeter mounted on a Heliopath stand, which requires considerably more operator technique, or to rely on the penetrometer for the higher viscosity ranges. Penetrometer data are considered to be less reliable because of the wide scatter of data points.

Reactivity or Cure

Previously, a measure of the reactivity of a resin mix was obtained by casting a sample in a test tube or small beaker placed in a hot bath and recording the exotherm developed. Of course the effect of the glass reinforcement, if any, is ignored. More recently an Audrey™ II dielectric analyzer (made by Tetrahedron Associates) has been used in conjunction with an X-Y recorder and a modified flat sheet mold. That technique determines the exact moment of cure and the rate of cure of a fully compounded molding compound. A flat sheet matched metal mold has been modified to include in each mold half a 2-inch square polished tool steel electrode mounted in the center of the mold surface and electrically isolated from the mold. Attached to each electrode is a coaxial cable that runs through channels drilled in the mold sections to the Audrey™ dielectric analyzer. As the mold closes, the two electrodes are opposite each other and parallel and separated by the test material. When a dipole is placed between those electrodes, a capacitor is formed. It is therefore possible to monitor the changing dielectric properties of the sample being molded as it heats up and cures. Figure 7.2 contains a schematic diagram of such a flat sheet mold construction.

Audrey™ has a capacitance of from 0 to 500 picofarad and can measure dissipation factor from 0 to 1 at frequencies of from 100 to 1,000 Hertz. On Audrey™ curves there is an initial heat-up period after which catalyst activation takes place. As the compound begins to cure, the capacitance swiftly decreases until the point at which complete cure is reached after which the curve is parallel, or nearly parallel, to the time axis. No other method for determining the cure of a fully compounded

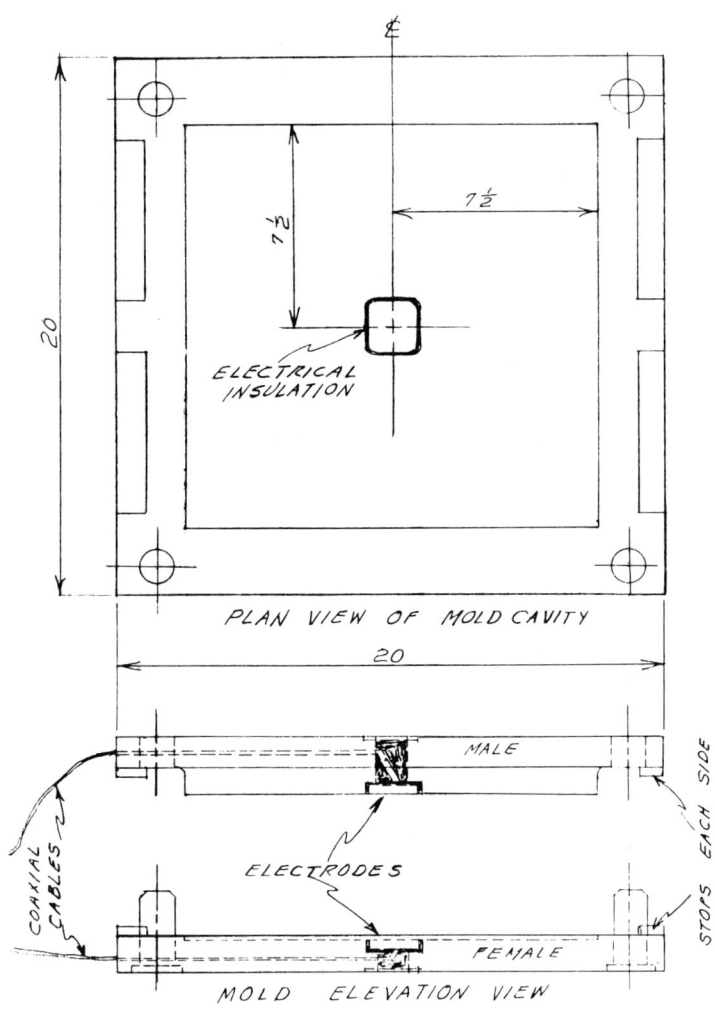

Figure 7.2 Mold Revisions for Dielectric Testing

molding material is available. The technique applies now only to flat sheets. In time it may be possible to include one electrode in a critical area of a production mold and to monitor the cure of the part. The use of bosses and ribs will complicate the interpretation of the results.

Spiral Flow

One of the more important parameters of molding compounds is its spiral flow length. Duralistic Products Company developed its Hydratec™ spiral flow test unit to measure that property under controlled conditions. A weighed charge is transfer molded into a spiral cavity. Charge weight, platen temperature, transfer pressure, and molding pressure may be varied. Figure 4.7 shows the type of material studies that can be made with that equipment. It is very easy to mold BMC in this unit. For SMC, however, it is necessary to cut the molding compound into ⅜-inch squares before preparing the charge. Otherwise the longer glass fibers plug the spiral flow path. Figure 7.3 shows one type of cutting template that has proven useful for preparing SMC charges for spiral flow testing. The template is placed on the SMC, and a knife is used to cut lengths ⅜-inch apart. The template is then rotated 90°, and the cutting is repeated to provide the compound squares.

Physical Properties of Molding Compounds

The physical properties of molding compounds are very important, and the manner in which the data are obtained has a direct bearing on the results. It is not enough to specify the ASTM or other test procedure to be used, but the manner in which the test coupons are obtained must be specified. Many laboratories have standard ASTM molds for preparing test specimen. For thermoplastic compounds injection molded specimen provide test data similar to coupons cut from medium-sized injection molded parts. When glass fibers are included in a molding compound formulation, the orientation of the glass in the final coupons has a direct bearing on the results obtained. If the glass is oriented in the longitudinal direction of the test spec-

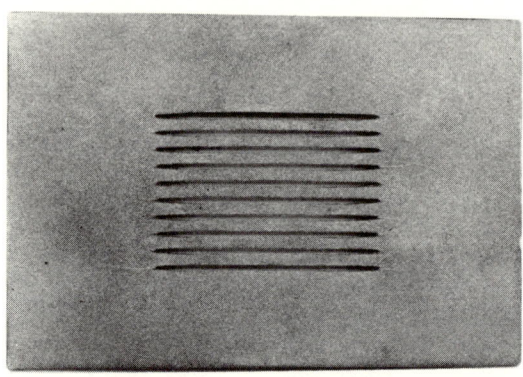

Figure 7.3 Cutting Template to Prepare SMC for Spiral Flow Testing

imen, misleading and unusually high test data may result. The most reliable test data are obtained from machining test coupons directly from molded parts. To minimize the effect of fiber orientation, test coupons should be cut both parallel to and perpendicular to the flow direction. For quality control of molding compound batches it is advisable to compression mold flat sheets and to machine coupons from those sheets. Figure 7.4 shows a comparison of compression-molded coupons versus coupons machined from a compression-molded flat sheet for a series of BMC compounds with glass contents of 10–30% by weight, keeping the glass length constant. The physical prop-

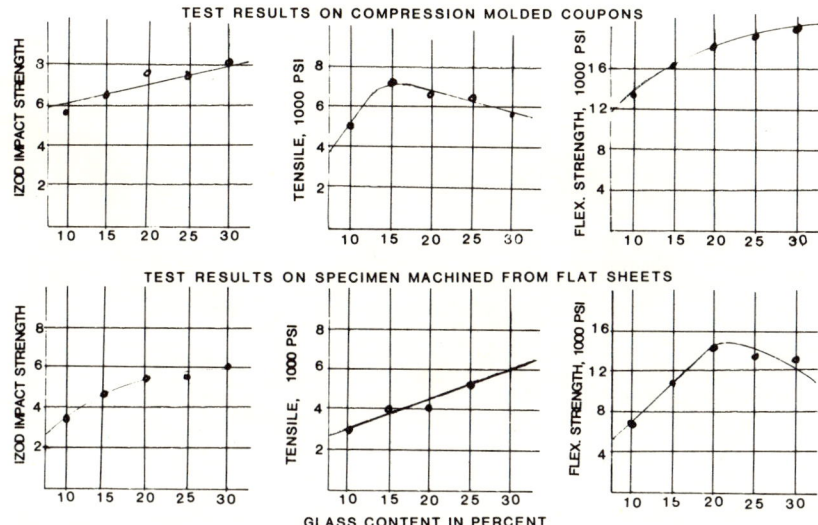

Figure 7.4 A Comparison of Test Results: Molded vs. Machined Specimen

erties tested increase with an increase in glass content but not necessarily linearly. As mentioned earlier, the level of physical properties is greater for the compression-molded coupons than for the coupons cut from flat sheets. Flexural strength is more sensitive to glass increase than tensile strength. As the glass content increases, it is necessary to increase mixing time in order to obtain equivalent glass wetting. Those higher values can be misleading. A definite pattern of coupon selection should be established and followed on all testing. Half the coupons should be selected parallel to a panel axis and the remainder perpendicular to this axis in order to check the effect of fiber orientation.

Molding Compound Flow

One of the important characteristics of BMC and SMC is their unusual ability to flow in the mold as heat and pressure are applied. Since the fiber lengths usually are different and the glass contents are different in BMC and SMC, one would expect

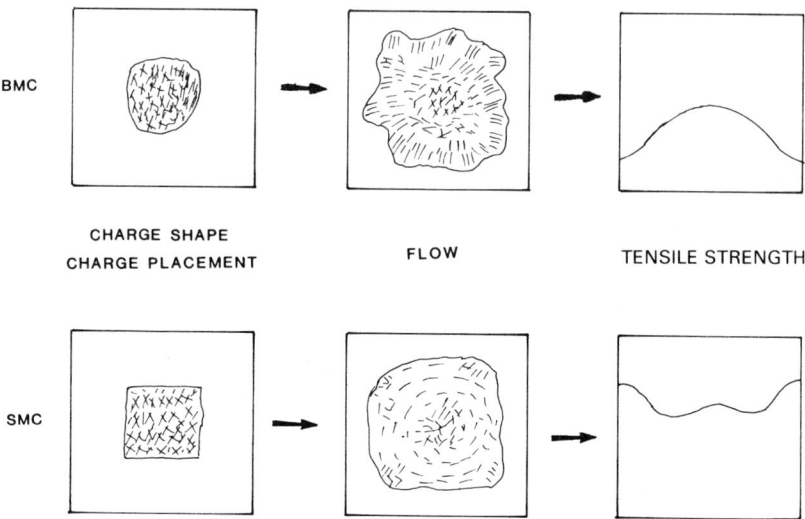

Figure 7.5 Fiber Orientation Owing to Flow

a somewhat different orientation as those two materials are molded. A study was made in a square flat sheet mold using the usual charge shape for each material with the charge placed in the center of the mold. Figure 7.5 is a sketch of the flow pattern and fiber orientation obtained by stopping the press platen before it rested on stops and allowing the charges to cure. Normally molded charges of each material were checked for tensile strength across the width of the panels as molded. A diagram of those results can be seen on the right-hand side of Figure 7.5. The BMC had a generally lower level of tensile strength than did the SMC, as would be expected. The edges farthest from the charge had the lowest tensile strengths. In the SMC panel the variation from one location to another across the panel width was not so great.

Figure 7.6 is a progressive sketch that shows the effect of molding compound flowing around an insert or core. A small void or crack can result on the downstream side.

Figure 7.7 is another progressive sketch that shows the effect of BMC flow when a mold has a leaky shear edge.

Figure 7.8 shows the effect of several mold surfaces on compound flow into a rib that is horizontal to the flow direction.

Compression Molding Technology 195

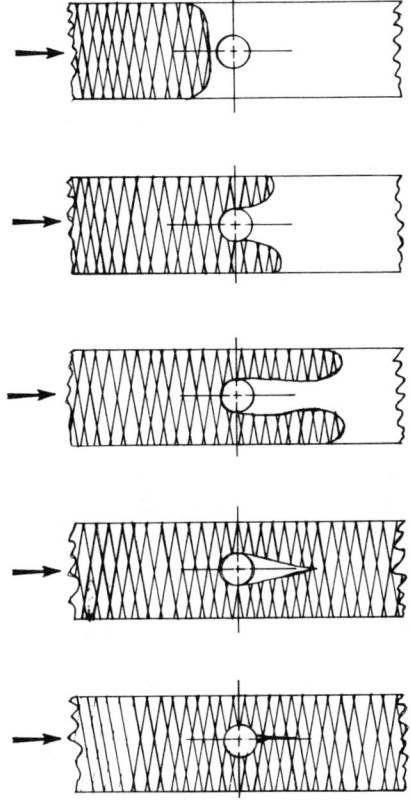

Figure 7.6 Flow Around an Insert or Core

Figure 7.9 is a set of progressive sketches that show how molding compounds flow into ribs. The top set illustrates how molding compound reacts when flowing perpendicular to a rib. The center portion is a view from above the rib, and the behavior is shown as the compound flows parallel with the rib and into the rib.

Preparing BMC Charges

The most popular way to consolidate a BMC charge in preparation for molding is to run it through an extruder to form a log. Flat and rectangular extrusions also have been prepared

196   Handbook of Polyester Molding Compounds and Technology

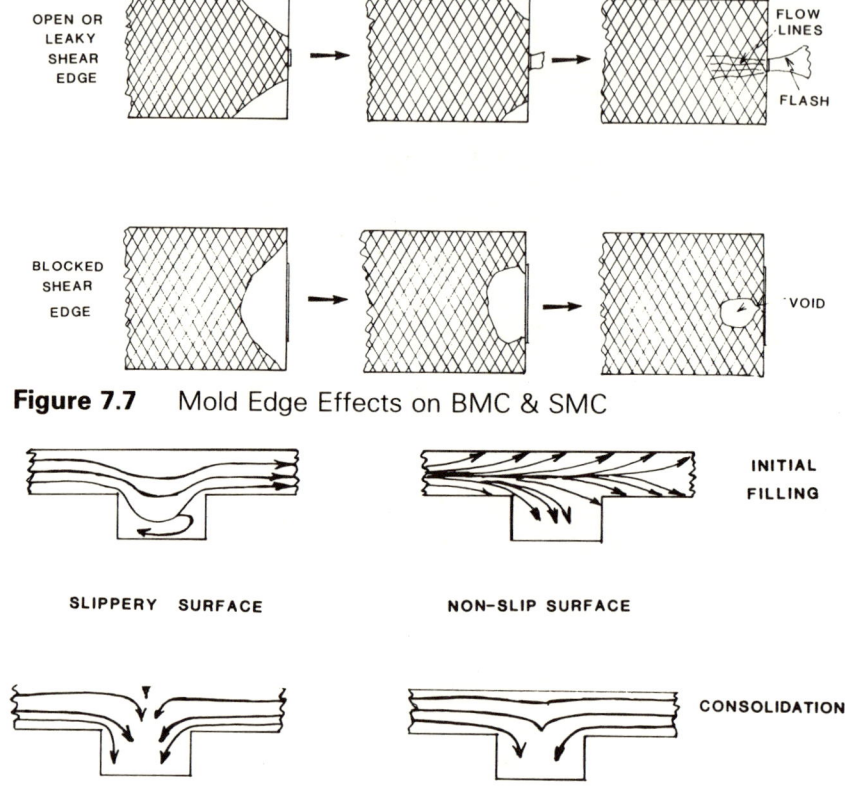

Figure 7.7   Mold Edge Effects on BMC & SMC

Figure 7.8   The Effect of Mold Surface Texture on Flow

for special parts. Bulk molding compounds containing more than 25% glass by weight generally cannot be extruded satisfactorily. Hand-shaped charges are frequently used, and sometimes charges can be compacted by placing them in a box die and compressing them with an air cylinder or mechanical press. It is important that the charges be the same part after part.

Preparing SMC Charges

Generally, rectangular patterns are the most economical to cut from SMC sheet, but oval, round, and special patterns can be die cut with clicker presses and steel rule dies or with a short

Compression Molding Technology   197

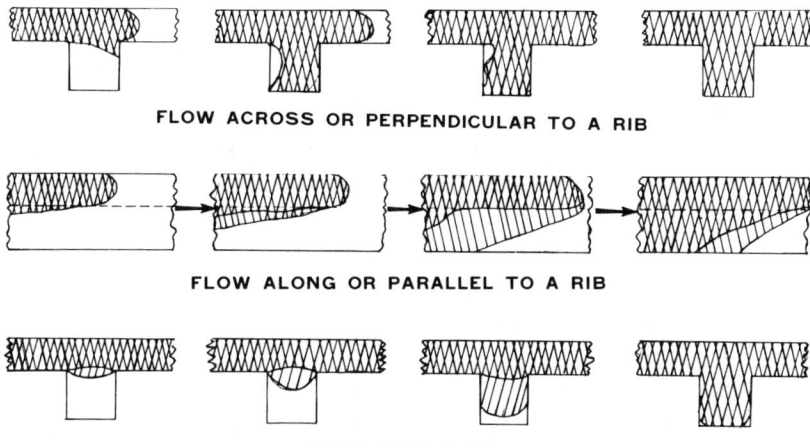

Figure 7.9   Several Methods of Filling Ribs During Flow

bladed knife and a template. The polyethylene film must be removed before placing the charge in the mold. If the part is thick, whereby several plies of SMC are needed to make a charge, the polyethylene film should be removed from all internal plies as they are stacked together, but left on the outside plies to minimize styrene evaporation until the charge is placed in the mold. Much operator time can be saved in the preparation of rectangular charges by having the sides trimmed in-line on the impregnator as the SMC is being made.

Breaking in New Molds

An external mold lubricant should be used on all new molds before their use and on all molds after they have been in storage. After the mold has been brought up to temperature, carnuba wax or zinc stearate is applied liberally to all mold surfaces. Excess lubricant can be blown off with an air hose. Generally, a mold break-in compound is used for the first few heats. When parts are releasing satisfactorily and after it has been determined that there are not undercuts in the mold that will mechanically interfere with the part releasing from the mold, the normal SMC compound should be used. If there are knock out pins in the mold, their travel should be adjusted so that they push gently against the cured part at the end of the press opening cycle. The mold is now ready for operation. The molding variables described earlier must then be considered and a judgment made on setting limits to produce the desired part.

Molding BMC and SMC

After limits have been established for all the molding variables on a new job, a molding specification sheet should be completed and should include a sketch of the mold charge shape and of the desired charge placement in the mold so that the setup can be duplicated on future runs. If one of the molding variables limits is changed for any reason, the molding specification sheet should be revised. When problems develop during production, their history should be documented for future reference. The troubleshooting guide in Table 7.4 can be useful in eliminating defects and production problems as they develop. While the guide specifically covers SMC, many of the problems and solutions are common to BMC as well.

SMC Troubleshooting Guide—Molding

Problem	Description	Possible Cause	Remedy
Mold Not Filled	The mold is not filled out	Charge weight too low Mold temperature too high. SMC cures before filling out mold	Increase charge weight Lower mold temperature
		Press closing time too long. SMC cures before filling out the mold	Reduce closing time
		Pressure too low Charge area too small	Increase molding pressure Increase charge area
	The mold remains unfilled at the edges only in a few places	Charge weight too low Charge escapes before mold closes	Increase charge weight Place charge more carefully
		Shear edge clearances too large, allowing SMC to escape before mold closes	Reduce shear edge clearances by welding or chrome plating
		SMC shear edge telescope too short, allowing SMC to escape	Increase telescope depth. If fault is minor, excess charge may help.
	The mold is not filled out in spots although the entire edge is filled	Charge weight too low Air cannot escape from the mold	Increase charge weight Rearrange charge so that air can be pushed ahead of the advancing fronts
		Blind pockets in mold prevent air from escaping	Install self-cleaning knockout pins in the mold If the fault is slight, increasing the pressure may help

TABLE 7.4 Continued

Problem	Description	Possible Cause	Remedy
Blisters	Round elevations on the surface of the cured part	Air entrapped between the SMC layers in a charge	Decrease charge area to force more flow to remove air Compress charge layers to expel air before placing in the mold
		Too high mold temperature (monomer vapors)	Lower mold temperature
		Dry glass	Lower resin mix viscosity Install more work rolls on SMC impregnator Heat work rolls on impregnator Change glass reinforcement
		Flash left in mold	Remove flash more carefully from the mold
		Too soft SMC	Mature SMC longer and/or increase thickening agent concentration
		Undercured part (monomer vapors)	Increase cure cycle and/or mold temperature
		Only on thick-walled areas Internal stress ruptures laminate	Decrease charge area to force more flow to mesh fibers better Reduce mold temp and/or increase cure cycle
		Weak spot along a knit line	Shorten flow path by increasing charge area

Compression Molding Technology

Blisters (cont'd)	Decrease in strength in one direction in places with long flow paths (glass orientation)	Shorten flow path by increasing charge area
	Area of charge too large. Air cannot escape because flow path is too short. Open shear edges on mold Dry glass Too high initial mix viscosity	Decrease charge area. Add small charge on top of larger charge Repair shear edges by welding or chrome plating Reduce initial mix viscosity
Porosity	Glass not sufficiently wet out	Install additional work rolls on impregnator Change glass reinforcement
	Pregelled area on a part	Lower mold temperature Add inhibitor to resin mix
	Dry-out of SMC charges	Wrap charges with a vapor barrier film Do not prepare charges too far in advance of molding
	Too soft SMC flow	Increase maturation by additional time and/or storing at elevated temp. before molding Increase thickener level
	Insufficient pressure High zinc stearate levels in SMC	Increase molding pressure Reformulate SMC reducing zinc stearate level

TABLE 7.4 Continued

Problem	Description	Possible Cause	Remedy
Diesel burning	Dark brown or black surface in places when part is not completely filled out	By compressing trapped air and vapors the temperature is raised to the ignition point	Choose a charge that will not trap air but pushes it ahead as SMC flows On blind holes or pockets, provide vents for the vapors to escape
Part sticking in mold	It is difficult to remove the cured part from the mold. In some spots the SMC sticks to the mold	Mold temperature too low Cure cycle too short SMC was unwrapped too long	Increase mold temperature Increase cure time Keep SMC sealed with a vapor barrier film until ready for use
		Mold not broken in. Mold is new or has not been used for some time	Use an external mold release on the first few moldings
	The cured part is difficult to remove. In some spots SMC sticks to mold. At same time pores and scars show on surface	Area of charge too large. Air on surface cannot escape because of short flow path. Trapped air delays cure	Decrease charge area. Add small charge on top of larger charge
Internal cracks	Not normally noticed unless the part is cut open, exposing the crack	Only with heavy walls, bosses, or ribs. The part cracks because of thermal stresses	Decrease charge area so that flowing glass fibers mesh together better Lower mold temperature

Warping	Part is slightly warped	Warpage owing to shrinkage during curing and cooling	Cool part in jig. Increase LP additive in formulation.
	Part is severely warped	One mold section much hotter than other selection. Warpage owing to glass fiber orientation.	Reduce temperature differential in mold sections. Short flow path by increasing charge area.

CHAPTER 8

Injection Molding Technology

General Information

Conventional injection molding is the most widely used process for making plastics products. In that process heat and work-softened thermoplastic molding compounds are forced into a mold, where they are hardened by cooling. The original machines used a plunger and heaters to soften and move the molten compound to the mold (Figure 8.1 is a schematic diagram of such a machine). Modern injection machines use an extrusion type screw to mechanically heat and plasticize the compound and then inject the material into the mold. Those units are called reciprocating screw machines. Molding pressures of 5,000 psi commonly are used in reciprocating screw machines, and the mold must be able to withstand that force applied over the projected area of the part. Figure 8.2 is a schematic diagram of a reciprocating screw machine showing the several positions of the screw.

Ejection pins are used to force the part from the cavity at the end of a cycle. A mold may have one or more cavities. Sprues and runners transport the molten plastic compound to the cavities. The design of injection molds has progressed to the point where product complexity is almost unlimited. Side operating cores permit molding in some undercuts. Inserts can be automatically loaded into the molds. The injection molding of thermosetting materials is more properly termed *automatic transfer*

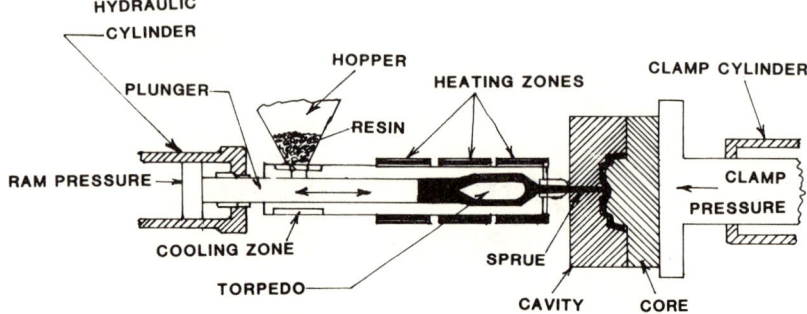

Figure 8.1 Schematic Diagram of a Typical Plunger Injection Molding Machine

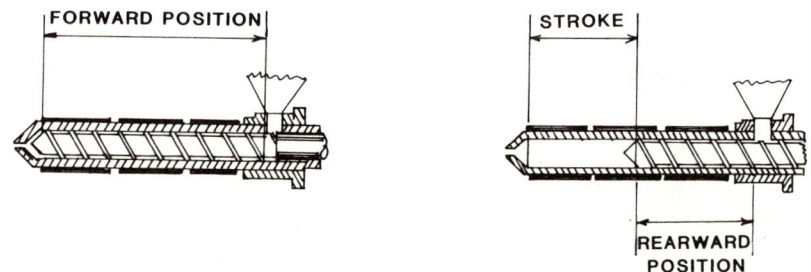

Figure 8.2 Reciprocating Screw Injection Molding Machine Showing Several Screw Positions

molding. Similar machines and mold design are used, but the molding compound is chemically hardened rather than hardened by heat transfer. In molding thermosetting compounds heat must be added to the compound to initiate chemical reaction rather than removing heat from the process. Figure 8.3 is a diagram of a typical injection mold with the nomenclature generally used to describe its component parts.

History of Thermosetting Injection Molding

Very early work in thermoset injection molding tended to use the standard molding equipment available at the time with the material being compounded into a dry granular form. Phenolic

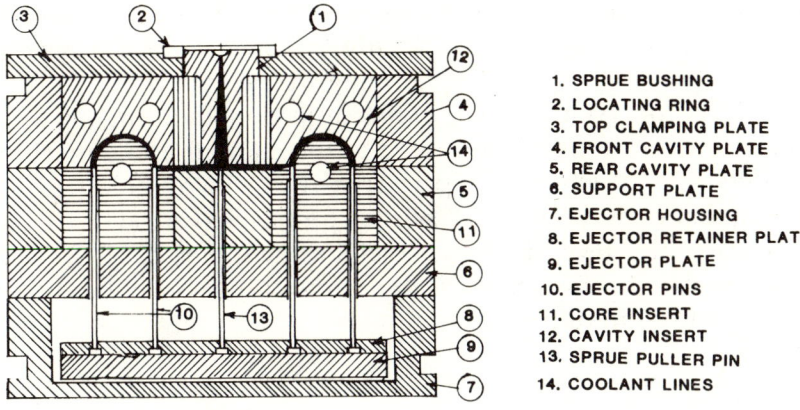

Figure 8.3 Typical two-Plate Injection Mold

resins were most often used, although some melamine and ureas were tried. However, using that approach meant that only very short glass fibers of ⅛ inch were possible. Longer glass lengths tended to clog the feed hoppers. Premix compounds generally were compression molded for large parts and transfer molded for small parts.

Rheinfrank[86] described the operation of an automatic rotating cold plunger machine for bulk molding compound (BMC) molding. The alkyd molding compound or BMC was first extruded into log form of approximately 3 inches diameter by 9 inches long in a rotary clay-type extruder. Those logs were charged to the vertical ferris-wheel-type loader on the horizontal injection machine.

The author had his first experience in injection molding BMC using a Seidl 6-station rotary injection molding machine that previously was used to mold rubber O rings. That work demonstrated the feasibility of injection molding BMC, but fast cycle times could not be evaluated due to the construction of the test molds. Building new molds was impractical since the six vertical press stations on the rotary Seidl machine were too small to mold a part large enough to be meaningful.

Hoffman[130] described the operation of the Instaset molding process using a pressurized cylinder to force the BMC into the closed heated mold through a nozzle that is heated during the material transfer and cooled during the balance of the cycle to prevent the BMC from polymerizing in the cylinder.

In 1968 the reciprocating screw injection molding of BMC was being very actively promoted by injection molding machine builders.[152,153,154] The author participated in several molding trials during that period. The most serious problems were (1) difficulty in loading the BMC into the injection cylinder, (2) fiber degradation due to the action of the screw, (3) severe fiber orientation in the molded part.

RAM Versus Screw Injection

During that period an experiment was performed on a 200-ton clamp pressure injection molding machine to determine the effect of the screw and ram on fiber degradation. A low profile BMC with approximately 16% glass content was used for the experiment. Large production batches of this BMC were prepared using ¼ inch glass lengths. Flat sheets of the virgin compound were molded in the usual manner. The balance of the compound was driven to the injection molding machine location, and a molding trial was made using a thermoset screw. Midway through the run some of the BMC was collected as it emerged from the nozzle (an air shot) and flat sheets of the sample were compression molded. That night the screw was replaced with a ram for a comparison molding trial the following morning. Again during the plunger injection molding trial an air shot was collected, and that material was compression molded into flat sheets.

The results of the trial will be found in Table 8.1. A map of the coupon locations in the 15 × 15-inch flat sheet is shown in Figure 8.4. Half of the coupons were cut parallel to the front of the press, which axis was arbitrarily labeled horizontal. The balance of the coupons were cut on a 90° axis to that, which was labeled vertical. By that procedure any effect of fiber orientation can be observed. When the average horizontal results

TABLE 8.1 Test Results on Screw Versus Plunger Injection Molding Trial (All coupons were machined from compression molded flat sheets)

	Virgin BMC	BMC Through Screw & Nozzle	BMC Through Ram & Nozzle
Notched Izod Impact Strength (Ft. Lb./In.)			
General Average (All Coupons)	4.25	1.74	3.71
Horizontal Average	4.74	1.96	3.69
Vertical Average	3.76	1.52	3.74
Orientation Index	1.26	1.29	0.99
Flexural Strength (psi)			
General Average	12,950	9,980	12.745
Horizontal Average	11,880	7,695	12,420
Vertical Average	14,020	12,270	13,075
Orientation Index	0.85	0.63	0.95
Flexural Modulus (psi $\times 10^6$)			
General Average	1.33	1.36	1.43
Horizontal Average	1.31	1.25	1.40
Vertical Average	1.36	1.41	1.47
Orientation Index	0.96	0.85	0.95
Tensile Strength (psi)			
General Average	5,440	5,700	5,600
Horizontal Average	5,770	5,660	5,710
Vertical Average	5,115	5,750	5,490
Orientation Index	1.13	0.98	1.04

NOTE: For a map of the location of all coupons see Figure 8.4.

are divided by the average vertical results, an index of orientation is obtained. Any values between 0.75 and 1.25 are considered to be satisfactory.

Table 8.1 shows that the screw caused a loss of 59% in impact strength while the ram loss was 13%. Similarly the screw had a loss of 23% in flexural strength and only a loss of 1.5% for the ram. That suggests, as one would expect, that there is far more fiber degradation caused by the screw than by the ram. Flexural modulus and tensile strength are not very much affected by glass degradation, and that showed in the results.

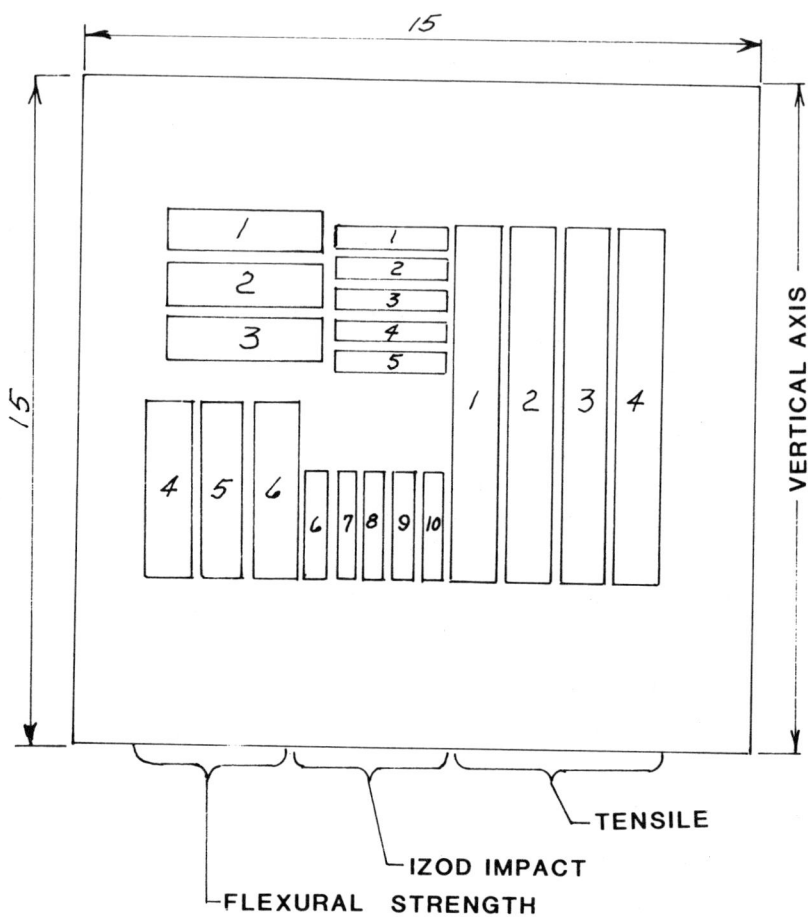

$$\text{ORIENTATION INDEX} = \frac{\text{HORIZONTAL AVERAGE}}{\text{VERTICAL AVERAGE}}$$

Figure 8.4 Location of Machined Test Coupons in a Compression Molded Flat Sheet

A similar study was run earlier using two different LP-BMC materials (those data will be found in Table 8.2). Again, compression molded flat sheets were prepared from the compounds after several process conditions. Coupons were cut according to the pattern shown in Figure 8.4. Spiral flow measurements also were run on the compounds using Hydra-Tec™ standard spiral flow equipment. On one compound, in addition to obtaining air shots, samples were removed from the stuffer at the end of the experiment. In spiral flow tests the glass reinforcement interferes with compound flow; the more fibers that are present the less is the flow. Conversely, as the fibers are destroyed in mixing or processing, the flow length increases. In those data it is evident that the stuffer had very little effect on compound A, the ram had a slight effect (9%), while the screw had a large effect (260%).

In impact strength the virgin value of compound A, the sample from the stuffer, and the ram-extruded sample all are approximately equivalent, while the screw-extruded sample lost 68%. Tensile strength and flexural modulus were affected more on that test than on that quoted previously. Flexural strength values showed a comparable decline with the screw being more injurious to the compound than was the ram.

Both the experiments showed that the ram or plunger did considerably less damage to the glass fiber reinforcement than does a screw.

Effect of Glass Lengths

Shortly after the Seidl-Battenfeld thermoset injection molding machine became available in the United States the author participated in a molding trial involving several different lengths of glass fiber strands in an 8½-inch diameter dish test mold. For that trial production batches of a LP-BMC were prepared; one batch contained ½-inch long glass reinforcement, and the other batch contained ¼-inch glass lengths. Coupons were cut

TABLE 8.2 Test Results on Screw Versus Plunger Injection Molding Trial

Sample Number		1	2	3	4	5	6
Material		A	A	A	A	B	B
Sample Conditioning		Virgin BMC Bulk Condition	BMC Removed from Stuffer after Being Compressed	BMC through Nozzle with Plunger	BMC through Nozzle with Screw	BMC through Nozzle with Plunger	BMC through Nozzle with Plunger
Spiral Flow Length in Inches	X	7.8	7.9	8.5	28.3	11.5	19.5
	R	6.5–8.8	7.8–8.3	8.0–9.5	28.0–28.9	10.8–12.5	16.5–21.8
Izod Impact Strength, Ft. Lb./In.	X	5.27	5.43	5.23	1.69	6.73	2.50
	R	3.33–6.60	3.92–6.83	4.50–6.26	1.42–2.09	5.26–7.77	1.59–3.42
Tensile Strength, psi	X	5,175	5,560	4,040	4,080	3,720	4,000
	R	4,050–6,525	3,500–6,680	3,500–4,705	3,280–4,470	3,260–4,270	2,550–4,670
Flexural Strength, psi	X	15,890	16,450	13,140	9,990	13,810	13,150
	R	14,150–17,010	14,720–17,420	9,890–16,450	6,930–11,570	10,470–17,100	11,320–15,030
Flexural Modulus, psi × 10^6	X	1.51	1.51	1.35	1.20	1.44	1.35
	R	1.35–1.63	1.40–1.58	1.16–1.45	0.94–1.43	1.32–1.50	1.24–1.47

NOTE: All coupons were machined from compression molded flat sheets. For a map of coupon locations see Figure 8.4.
X = Average Value
R = Range of Values

TABLE 8.3 Physical Properties Using the Seidl-Battenfield Plunger Injection Machine

¼-in. Length Fiberglass Reinforcement

Property Tested (Average Value)	Compression-Molded Flat Sheet	Injection-Molded Dish	Air Shot Compression Molded
Impact Strength Ft-Lb./In.	4.5	5.6	3.9
Flexural Strength psi	16,160	12,060	14,540
Flexural Modulus psi	1.63×10^6	1.23×10^6	1.58×10^6
Tensile Strength psi	6,240	b	5,020

½-In. Length Fiberglass Reinforcement

Property Tested (Average Value)	Compression-Molded Flat Sheet	Injection-Molded Dish	Air Shot Compression-Molded[a]
Impact Strength Ft. Lbs./In notch	3.5	3.5	3.3
Flexural Strength psi	12,780	10,160	11,240
Flexural Modulus psi	1.60×10^6	1.17×10^6	1.59×10^6
Tensile Strength psi	4,840	b	5,210

a. Air shot made through small (¼ in. dia.) nozzle
b. Dish not large enough for test specimen to be cut

from the molded dishes in two directions similar to the pattern shown in Figure 8.4 but on a much smaller scale. Virgin BMC from each batch and air shots were collected that were compression molded into flat sheets for testing. The same coupon layout as shown in Figure 8.4 was selected. Table 8.3 contains the test results from that evaluation.

Physical properties on the compression-molded virgin BMC containing ¼-inch glass lengths are considerably better than for the identical compound with ½-inch glass. That is attributed to the better mixing action afforded by the shorter glass lengths, and that phenomenon has been observed previously. There was no degradation shown on impact strength on the injection molded dishes with ½-inch glass and an improvement in dishes containing ¼-inch glass. That probably is due to the fiber orientation being aligned parallel to the direction in which two of the three coupons were cut. During injection the flow on that dish approximates concentric circles around the center sprue location. In that case fiber orientation is working to improve physical properties. As is to be expected, the air shots show slight damage because of the action of the compound being forced through the nozzle.

In this study the ¼-inch glass length appears to be optimum for injection molding.

Injection Molding Process Recommendations

Since the molding compound is forced into a closed hot mold, the problem of precure in compression molding caused by the charge lying on the hot open mold too long as the press closes is completely eliminated; therefore, a much hotter mold temperature is permissible, which greatly reduces cure time. The automatic injection of compound and ejection of cured parts also speeds up the overall injection molding cycle. Faster catalyst systems generally are employed to further minimize cure times.

Figure 8.5 consists of a sketch of a hypothetical box and the normal process conditions that would be used to mold the box by compression molding and by injection molding using a good BMC material.

Runner and Gate Design

The largest internal diameter standard sprue bushing that can be tolerated should be used in order to minimize fiber degradation during the injection cycle. The ¹¹⁄₃₂-inch diameter bushings are recommended whenever possible.

Injection Molding Technology 215

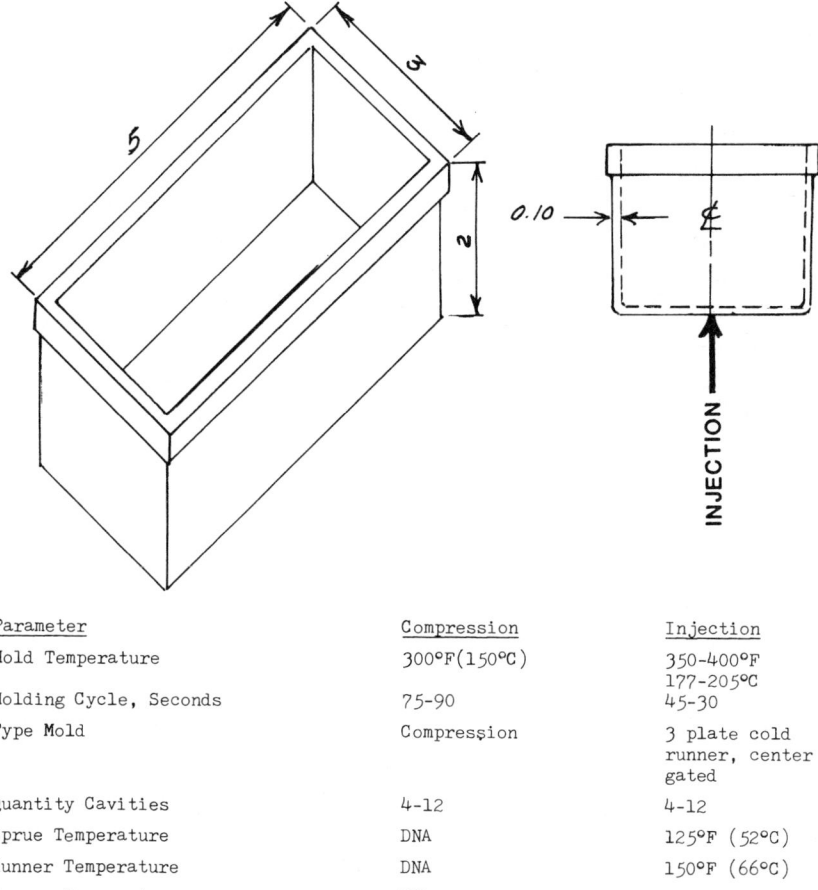

Parameter	Compression	Injection
Mold Temperature	300°F (150°C)	350-400°F 177-205°C
Molding Cycle, Seconds	75-90	45-30
Type Mold	Compression	3 plate cold runner, center gated
Quantity Cavities	4-12	4-12
Sprue Temperature	DNA	125°F (52°C)
Runner Temperature	DNA	150°F (66°C)
Hopper Temperature	DNA	Room Temp.
Catalyst System	1% TBPB BOR	1% TBPO BOR
Minimum Clamping Pressute, Tons	DNA	150-450

BOR=Based on resin. DNA=Does not apply. TBPB=t-Butyl Perbenzoate
TBPO=t-Butyl Peroctoate.

Figure 8.5 Comparison of Process Conditions for BMC Molding by Compression and by Injection Molding

Runner and gate design of injection molds for thermosetting compounds are very important factors in obtaining fast and uniform cavity fill without gas entrapment and/or precure.

Three basic types of runners generally are used: rectangular, half-round, and round. Rectangular runners are easier to machine into the die plate but are more difficult to polish. Half-round runners are milled into one die plate and are easy to polish. For full round runners a semicircular groove is machined into each half of the die plate. Both grooves must match exactly to permit easy material flow. That type runner provides fastest fill times with the least injection pressure. Vaill[152] gives a good discussion of the design of runners and gates.

Espenshade[201] gives a good discussion of the various types of gates and their effect on material flow and physical properties and of the different locations on a laboratory flat sheet molded part. Included is a study of the effect of glass type on physical properties, the effect of several different screw tips on the physical properties of molded parts, and the effect of gate and runner systems on physical properties, etc.

Troubleshooting Molding Problems

Figure 8.6 consists of a chart that might be useful in troubleshooting thermoset injection molding problems. Across the top is a list of 19 problems that occur from time to time. Possible remedies are listed in the vertical column on the left of the chart. Several possible remedies are listed for most problems. Some of the remedies are easy to evaluate and can be completed in a minimum time; others require considerable effort and fairly long runs to evaluate their effectiveness. It is wiser to investigate simple machine variables first before scheduling mold changes such as enlarging sprues or improve mold venting as listed under "cavity not filled." The latter changes may be required in extreme cases, but they require considerable mold downtime to complete.

Figure 8.7 shows a 12-cavity mold for electrical outlet boxes about a second after the ejectors have pushed the molded premix off the male mold. Injection molding is an extremely useful process for such small parts and is employed by several molders to make these and similar items.

Injection Molding Technology 217

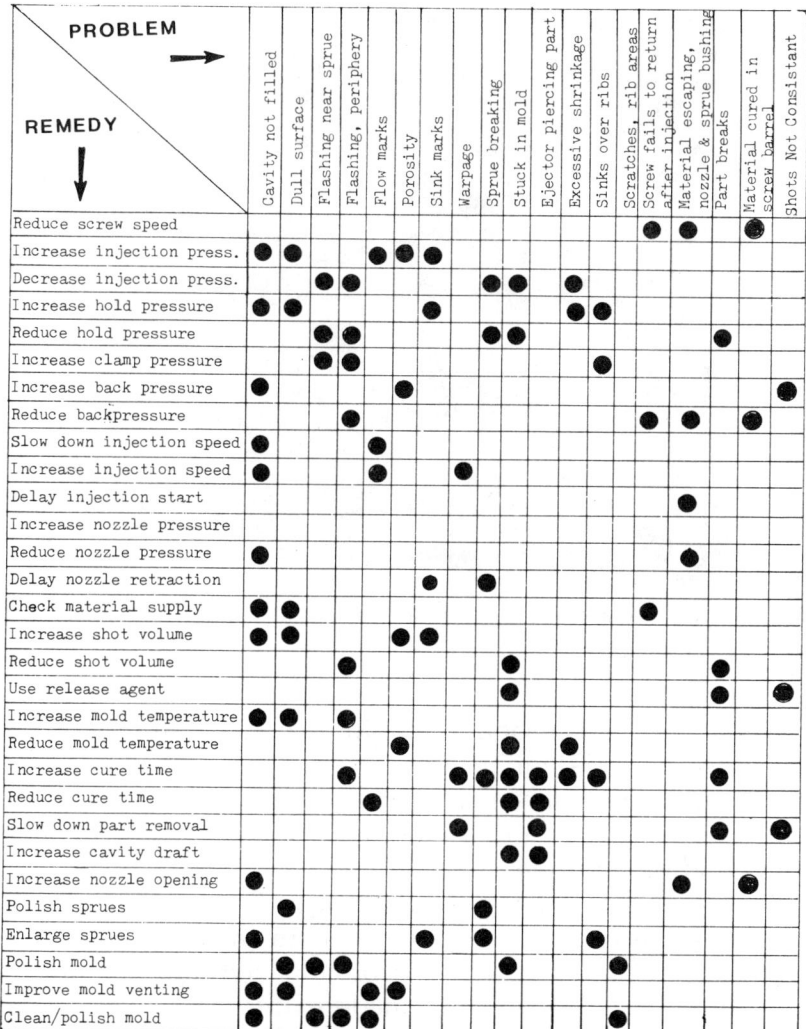

Figure 8.6 BMC Injection Molding Troubleshooting Guide

Continuous Mixing/Molding System

Rizzi[256] discussed a machine and process developed by The Farrel Company Division of USM Corporation that it termed its *Total System* approach to automatically and continuously

Figure 8.7 A Multicavity Production Mold Automatically Ejecting Parts (*Courtesy: Allied Moulded Products*)

mix, meter, and injection mold FRP materials. The major components of the system include continuous metering-type feeders, a continuous mixer, a continuous premix device to feed the injection screw, and a modified reciprocating screw injection molding machine.

That system was later modified so that it had the capability of continuous or batch operation or both and a variable mixing intensity to control the amount of work put into the mix. The new unit was called the Farrel medium intensity mixer (FMIM). It was described in an article in *Plastics Techonology* (May 1975).

FRP Injection Molding Markets

Table 8.4 is an estimate of the FRP injection molding shipments for the years 1976 and 1983 for both thermoplastic and thermoset materials broken down by SPI market areas.

TABLE 8.4 Estimates of FRP Injection Molding Shipments by Markets (in millions of pounds of composite)

Family	Thermoplastics		Thermosets	
Year	1976a	1983b	1976a	1983b
Aircraft & Aerospace	—	—	—	—
Appliance & Bus. Equip.	24	41	10	12
Construction	—	—	—	—
Corrosion-Resistant	—	3	—	—
Electrical & Electronic	32	54	15	21
Marine	—	—	—	—
Land Transportation	89	198	—	27
Consumer	3	6	—	—
Miscellaneous	2	22	—	—
Total	150	304	25	60

Source: aSPI RP/CI; Includes both thermoplastic & thermoset composites.
bUnofficial estimate.

No applications have been found for either material family in the aircraft, construction, and marine areas. Very nominal quantities of reinforced thermoplastics are being used in the corrosion, consumer, and miscellaneous market areas while no thermosets are being used in those three areas.

Injection molded reinforced thermoplastics have their largest market area in land transportation, mostly in under-the-hood applications, while the electrical market is the second-largest area. The use of reinforced thermoplastics more than doubled during the seven-year period covered by the estimates.

Injection molded reinforced thermosets increased their market penetration 240% from 1976 to 1983. Despite that achievement their total shipments amounted to only 20% of the thermoplastic shipments. In 1976 injection molded thermosets were used only in two market areas, business equipment and electrical applications, and those in only nominal quantities. By 1983 the two market areas increased only very slightly, but a completely new market in land transportation developed that is larger than either of the other two markets. Most of the new applications are under-the-hood small parts. Housings for small electrical hand tools continued to be an important market.

CHAPTER 9

FRP Painting

Introduction

This chapter is intended only to provide a basic grasp of the fundamentals of painting fiberglass reinforced plastic (FRP) parts and is not intended to be an all-inclusive course on the subject.

There is a considerable difference in satisfactorily painting a low profile bulk molding compound (BMC) or sheet molding compound (SMC) part today than in painting the original mat or preform and premix part of 25 years ago. The very rough FRP surfaces previously obtained required only that the internal mold release agent and contaminants be removed in order to obtain good adhesion of the primer coat. Obtaining a smooth surfaced part was another problem. The author well remembers attending meetings in 1965 where Corvette body engineers were attempting to have the surface finish specification of body panels changed from 1,200 to 800 microinches. Today those parts are molded with in-mold coated SMC, and surface finishes of 100 microinches are standard. Sheet molding compound automotive front ends have surfaces of 150–200 microinches while BMC parts can be as smooth as 100 microinches. Those points should be kept in mind while one reads the brief history of painted FRP parts that appears later.

Paints

Paint is a pigmented* film-forming material. Unpigmented paints that yield nearly colorless films are called varnishes. Automotive

---
*The term pigment in this chapter follows the customary paint usage, which includes colorants as well as items such as calcium carbonate, clays, aluminum trihydrate, etc., which are referred to as "fillers" in FRP usage.

paints most often are used to protect and to decorate. Over metal substrates paint retards corrosion and general weathering. That corrosion protection function is not required for FRP parts. Other special functions coatings provide are to reflect light, conduct electricity, insulate, etc. Table 9.1 consists of a classification of paints on the basis of their function and their application or curing characteristics.

Paint Components

In general most paints contain the following: (1) a resinous binder, (2) a solvent, (3) pigment(s), and (4) additives. The resinous binder is the solid material that forms the bulk of the paint film. It generally is a tough, amorphous, polymeric material that gives the paint most of its thermal, mechanical, chemical, and weathering properties. Table 9.2 compares the general properties of some common paint binders.

A solvent is a pure or mixed liquid that is used to liquify the paint before application. In practice the terms solvent, solvent blend, and thinner are often used interchangeably. Table 9.3 contains a list of solvent terms and their meanings.

Paint pigments are small solid particles present in the paint film to provide specific properties. Pigment particles range in size from less than 1 micron to more than 100 microns. They may be spherical (calcium carbonate), flat (china clay), or needle shaped (talc). The major reasons for using paint pigments are given in Table 9.4. A grinding operation is used by the paint manufacturer to disperse pigments in a vehicle. An important feature of pigment dispersion is the pigment volume concentration (PVC). The PVC is defined as the volume percent of the dry film that actually is pigment as opposed to binder or additives. PVCs range from zero for unpigmented films to more than 90% for zinc-rich primers. Figure 9.2 shows a schematic of paint films (greatly enlarged) that are underpigmented, fully pigmented, and overpigmented. The PVC at which the paint

TABLE 9.1 Paints Classified According to Their Function and/or Application Characteristics

Topcoat	The final coat. The actual film the eye sees.
Undercoat	Any paint film beneath the top coat.
Basecoat	The colorcoat under a clear topcoat.
Primer	An undercoat used to improve adhesion to the topcoat on FRP parts and to resist corrosion on metals.
Sealers	Usually an undercoat that improves adhesion between 2 paint films. Sometimes a sealer coat is applied to prevent movement of pigments, film formers, or oils.
Surfacer	An undercoat designed to fill surface imperfections. Surfacers are compounded for easy sanding.
Air Drying	Paints that do not need to be heated to cure. House paints and some automotive repair paints are examples.
Baking	Paint that is designed to be heated to affect cure.
Two-Part Paints	Paints having 2 reactive components that must be kept separate before application. The components are mixed immediately before use.
Lacquers	Paints that dry by evaporation of the solvent. No chemical reaction takes place in the paint.
Enamels	Usually paints that involve a chemical reaction between binder materials to make a solid film.
Latex	A paint applied as a water base emulsion. Drying involves evaporating the water and coalescence of the paint particles.
Powder Coats	Finely ground powder paints that are applied to a part and then heated to the melting point where the paint flows out to form a smooth film. This film solidifies on cooling.

particles touch each other (the space between is filled with binder) is called the critical pigment volume concentration or (CPVC). The quantity of pigment required to achieve CPVC depends on the size and shape of the pigment particles. Smaller particles can pack together with greater efficiency, which results in higher CPVCs. The significance of CPVC is shown in

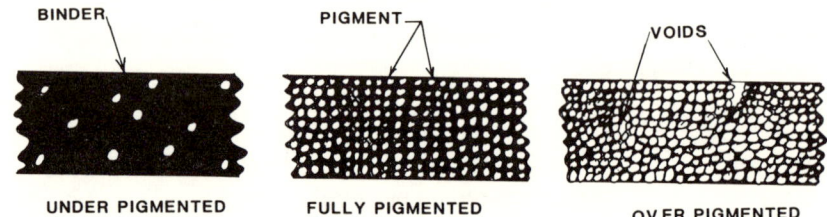

Figure 9.1   Three Possible Pigment Binder Ratios

TABLE 9.2   Comparative Properties of Some Common Paint Binders

Family	Moisture Resistance	Chip Resistance	Temperature Resistance	Material Cost
Acrylics	medium	low	medium	medium
Alkyds	medium	medium	medium	medium
Asphaltics	high	high	low	low
Epoxies	high	high	high	high
Polyesters	medium	medium	medium	medium
Urethanes	medium	high	variable	high

Table 9.5, which lists how a number of important film properties vary with pigment level. The actual pigment level in any paint generally is a compromise between several of the properties listed in this table.

Types of Paints

Table 9.6 consists of a classification that covers virtually all chemical coatings in use today. Until quite recently most industrial chemical coatings were of the low-solids content solvent-system type. As late as 1970 more than 95% of all industrial coatings in the United States were formulated as solvent-based systems containing less than 70% volume solids with actual volume solids probably averaging under 40%.[467]

Los Angeles Rule 66, which has now been adopted by many other areas, restricts the amount of hydrocarbon solvents that may be discharged to the atmosphere. The Organization of

TABLE 9.3 Definitions of Solvent Terms

True Solvent	A liquid that can dissolve the binder.
Diluent	A liquid that cannot dissolve the binder by itself but can be added to a solution to increase its capacity for the binder.
Latent Solvent	A liquid that cannot dissolve the binder by itself but increases the binder's tolerance for a diluent.
Thinner	Any pure or mixed liquid added to a paint to reduce its viscosity.
Front End Solvent	A fast-evaporating solvent that leaves the paint film very soon after application, usually before the paint reaches the oven.
Middle Solvent	A medium-evaporating rate solvent that leaves the paint primarily during flash-off and during warm-up in the oven.
Tail End Solvent	A slow-evaporating solvent that leaves the paint during the baking cycle.
Exempt Solvent	A solvent that does not react with sunlight to form smog compounds. Their use is often specified by legislation (such as Los Angeles Rule 66).
Reflow Solvent	A very slow-evaporating solvent that leaves the paint film during the thermal reflow stage of an acrylic lacquer bake cycle.
Retarder	A slowly evaporating solvent added to a paint to prolong drying time. It is used typically to reduce orange peel or blushing.
Solvent Blend	The particular mixture of liquids that gives a paint solution the desired properties. Skillful choice of the solvent blend allows a paint shop to adjust paints for local variations in temperature, line speed, humidity, application equipment, etc.
Evaporation Rate	The solvent must evaporate within the time allowed by the curing cycle considering available oven length, temperature, and production rates.
Mist Coat Thinner	A fast-evaporating solvent added to a paint to reduce its drying time.

TABLE 9.4 Reasons for Using Paint Pigments

Hiding Power	The ability to cover and hide from sight the coated surface.
Color	A specific color may be required to satisfy the customer's requirements.
Corrosion Resistance	Some pigments reduce the corrosion of metallic substances.
Mechanical Properties	Pigments often can increase the strength of a film or change its impact resistance.
Adhesion	This important property of a paint film can sometimes be improved by the proper selection of pigments.
Viscosity	The fluidity or liquidity of a paint is often related to the quantity and shape of its pigment particles.

TABLE 9.5 Usual Variations of Paint Properties with Increasing Pigment Levels

Property	Usual Effect of More Pigment
Viscosity	Increases
Adhesion	Maximum at CPVC
Tensile Strength	Maximum at CPVC
Moisture Permeability	Minimum at CPVC
Cost	Increases
Gloss	Decreases
Impact Resistance	Maximum at CPVC

CPVC = Critical Pigment Volume Concentration

Petroleum Exporting Countries (OPEC) oil embargo of 1973, which drastically increased the cost of energy, also had a significant effect on industrial coatings. Those two factors resulted in the quantity of low-solids, solvent-based paint systems being considerably reduced in favor of the higher-solids solvent systems, the two-component catalyzed systems, and the water-soluble or collodial dispersion systems. Most automotive assembly plants in the United States have switched to the electrocoat system for prime coats. That system presents unique problems for FRP substrates, as will be discussed later.

TABLE 9.6 Chemical Coatings Technology Types

1. Solvent systems under 70% solids by volume
2. Solvent systems over 70% solids
3. Two-part catalyzed systems
4. Emulsions and lattices
5. Water-soluble and collodial dispersions (including electrocoat)
6. Powder coatings
7. Radiation-cured systems
8. Vapor cure systems

FRP Molded Part Requirements for Painting

If one assumes that the FRP part has been properly designed and has been molded of an optimized formulation in a well-constructed, polished, chrome-plated hardened steel mold, there still are precautionary measures that should be exercised in order to apply a coating at minimal cost. All FRP parts must be clean and thoroughly dry before paint is applied. A problem area is static attraction of dust particles to the molded part surface. Another potential problem area is the relatively soft surfaces, as compared with metals, that can easily be damaged if good handling and packaging procedures are not followed. It is advisable to 100% inspect all FRP parts and to repair any defects before hanging those parts on a paint line. Many problems on painted parts are difficult to observe until after the part emerges from the paint ovens, after much time and materials have been expended. The thermal conductivity of FRP parts is considerably less than for metal parts. The part surfaces generally do not heat up as fast in forced air ovens, and the final part temperature generally does not reach oven-ambient temperature unless an exceptionally long residence time is employed. The final part temperature should be confirmed with a traveling thermocouple taped to a part.

Steel parts are usually given a bonderite or phosphate treatment for corrosion protection and paint adhesion before the finishing operation. That pretreatment is not required on FRP parts.

Surface Preparation

Fiberglass reinforced plastic part surfaces contain dirt, which is attracted because of the part static charge, oil from the operator's hands, oil and grease from the presses and overhead conveyors, and internal and external mold release agents. All of the contaminants will interfere with primer adhesion unless they are effectively removed.

The most effective method, so far devised, is to wash the part with solvent, such as acetone, isopropanol, or toluene as the part is hung on the paint line. The solvent kills the static charge and removes most contaminants. Care must be taken to use soft, clean rags and small containers of solvent and to change solvent and rags frequently. That procedure is quite operator dependent and can be the source of further contamination unless the rags and solvent are frequently changed (of course that adds to the part painting cost considerably). Taking such solvents on a paint line also is generally not permitted by most fire codes and insurance companies because of the fire hazard.

The solution generally used, where large volumes of FRP parts are involved, is to use a power washer. Figure 9.2 is a block diagram showing the power washer process details of one unit installed at a molder's plant to clean automotive FRP parts before painting. After start-up, considerable problems were encountered with part blisters. A study of the problem was made that determined:

1. The power washer, as originally used, was less effective in reducing water blisters than an acetone wipe with a clean rag.
2. Carryover of the detergent in pockets of parts was a serious source of contamination. If any water droplets are left

FRP Painting   229

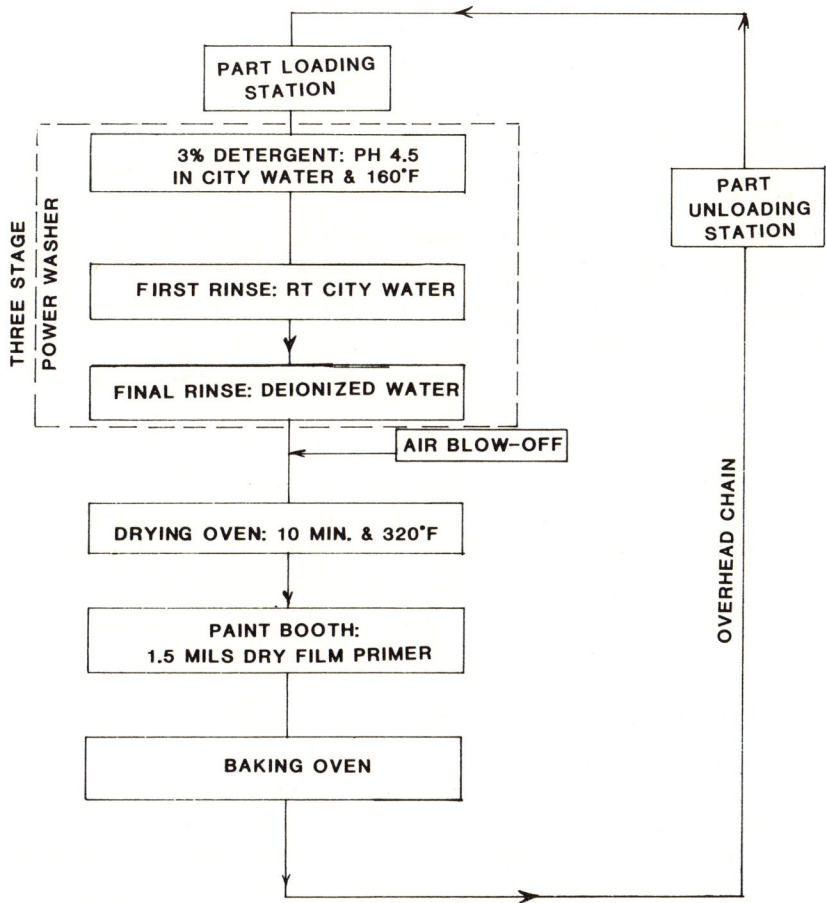

Figure 9.2 Typical Power Washer Process

on the part after the last deionized water rinse before drying in the oven, a high incidence of water blisters forms.
3. Water dripping from the chains and hangers, even in the drier, can cause water blisters.

It was later learned that even hot, deionized water leaches materials from the substrate of some formulations and leaves salt deposits on the part surfaces when the water evaporates. Those salt deposits can cause blisters.

A program was undertaken to learn how to satisfactorily operate the power washer efficiently. Another nearby FRP molder was operating a power washer on his parts satisfactorily. In discussions it was learned that this molder also had problems when he first started his power washer. He was conveying his parts through a deionized water high-velocity spray at 160°F (71°C), and no detergent or other dissolved ingredients were used. The FRP parts passed from the washer into an adjoining high-velocity air stream at 315°F (157°C) to blow all the water droplets off the parts and to dry the remaining water film. Parts were in the drier 3 minutes. It was further learned that when that molder first started his power washer, he used a phosphate detergent and normal air velocity. Under those conditions he experienced painting problems. The culprit was believed to be the detergent, but when it was removed from the process and hot city water was substituted, the blisters remained. Next, hot deionized water was tried, with some improvement, but that did not completely solve the problem. When high-velocity air jets were installed, the water blister problem was solved.

The molder's power washer was revised to include better part hangers, hoods to prevent water from dripping on parts on the conveyor line, and high-velocity air jets to strip all water from the parts before they entered the drying oven. The blister problem was solved. That line continues in operation today.

Automotive Primers

The principle function of a primer on FRP parts is to provide adhesion between the substrate and the top coat. Selection of the primer is dependent upon the final paint properties of the end product. In the automotive area, key properties are determined by standard test procedures such as: (1) tape adhesion test, (2) effect of humidity exposure; and (3) effect of gravelometer exposure. Basically primers are of the lacquer or enamel types. Lacquer (thermoplastic) primers allow time and temperature of the oven baking cycles to be varied over a wide range. Enamel (thermoset) primers require longer time periods at higher temperatures to cure. Certain precautions must be taken when one is using various combinations of top coats and primers. Thermoset topcoat solvents will attack a thermoplastic primer if used together. Thermoplastic top coats are compatible with cured thermoset primers. Normally primer films are applied to a 0.75–1.25 mil dry film thickness, which requires at least two separate spray applications, since a single pass of spray primer can be expected to deposit or build only 0.5–0.75 mil dry film thickness. Figure 9.3 consists of a block diagram showing a typical FRP prime coat system. Primer coats on SMC substrates should be baked at a higher temperature than the final bake temperature of the top coat. That will drive out solvents and volatiles that otherwise might be emitted during the final bake cycle. Some painting systems now subject the substrate, before painting, to a high temperature bake to remove any gases that might be present before applying prime coats.

Automotive Sealers

Sealer coats are paint films applied between a primer and a topcoat to improve adhesion. That is often necessary because the primer and topcoat may have quite different thermal expansion or impact-resistance properties. When the interface between the films is stressed, the layers sometimes separate and the topcoat can chip or break off. Sealers also may be used to provide a barrier coat between one paint film and the solvent in a subsequent coat. That may be necessary if the pigment in the lower coat has a tendency to migrate or bleed into the solvent of the upper coat. Sealers, like primers, are of the thermoset or thermoplastic type. They must be used in proper combination with the primer; for example, enamel (thermoset) sealers will attack lacquer (thermoplastic) primers and should not be used in this combination. Usable combinations include:

232 Handbook of Polyester Molding Compounds and Technology

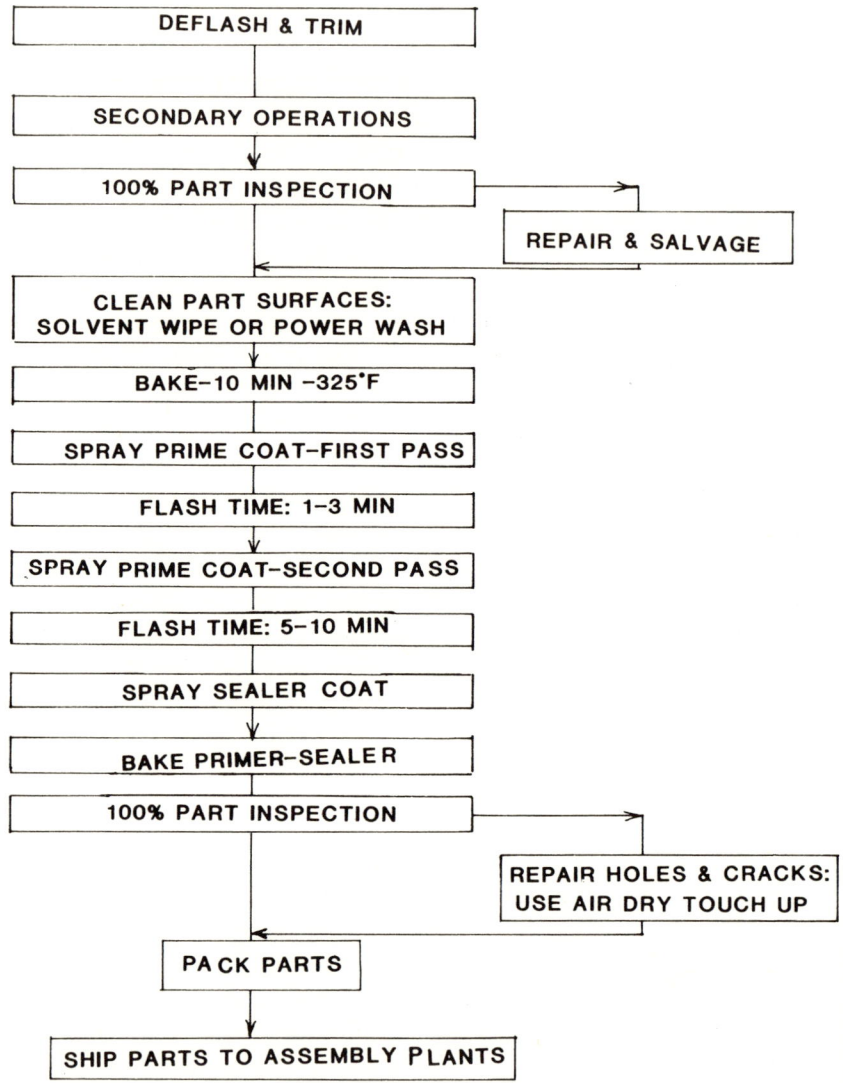

Figure 9.3 Typical Finishing Sequence for Prime Coated FRP Parts

1. lacquer primer, lacquer sealer, lacquer, topcoat;
2. enamel primer, lacquer sealer, lacquer top coat;
3. enamel primer, enamel sealer, enamel topcoat.

Sealers can be applied by the same spray equipment used in the finish operations. Usually the dry film thickness should fall in the range of 0.2–0.5 mils. Unlike primers, which are frequently applied to FRP parts by the molder before shipment to the assembly parts, the sealer coat is generally applied just before the topcoat application.

Sometimes sealer coats are pigmented a color contrasting with the primer coat so as to be used as a sanding guide.

Automotive Topcoats

The final paint film on the automobile, the one the eye sees, is the topcoat. It may be applied as one of several separate coats. If the paint is not baked, but only allowed to flash off, before receiving the next coat the process is known as a wet-on-wet application. The topcoat must supply the exterior surface with the desired color, gloss, abrasion resistance, and weatherability. Sheet molding compound moldings readily accept a number of standard topcoat paints. Basically, the most common finishing systems can be classified in one of the following: (1) lacquer-reflow (thermoplastic), (2) enamel-reflow (thermoset), and (3) enamel-normal (thermoset). Table 9.7 consists of the step-by-step procedures used to topcoat FRP parts at some automotive assembly plants. Sometimes the topcoat is applied to the FRP part after it has been assembled to the steel car, and all units are painted simultaneously. Frequently, however, the FRP part is painted on a separate line either before or after it is included in a subassembly, and then that subassembly joins the car later on the assembly line. Less floor space is required when the car is painted as a unit, and that is the current trend. Sheet molding compounds do not present unusual problems in a finishing operation; however, there are critical areas in the process that must be carefully considered to ensure a quality

TABLE 9.7 Top Coating Procedures For FRP Parts

Procedure	Lacquer-Reflow	Enamel-Reflow	Enamel-Normal
1. Clean All Parts	—	—	—
2. Apply Sealer Coat	0.2–0.5 mils (.005–0.13 mm)	N/A	N/A
3. Flash	1–2 min. @ booth temp		
4. Top Color Coat; First Pass			
5. Flash	1–2 min. @ booth temp.	1–2 min. @ booth temp.	1–2 min. @ booth temp.
6. Top Color Coat; Second Pass.			
7. Flash	1–2 min. @ booth temp.	1–2 min. @ booth temp.	1–2 min. @ booth temp.
8. Color Coat; Third Pass*	1.8–2.5 mils	1.5–1.8 mils	1.5–1.8 mils
9. Final flash	10 min @ booth temp.	3–5 min. @ booth temp.	3–5 min. @ booth temp.
10. Prebake Cycle**	7 min. minimum @ 180–200°F (82–93°C)	7–10 min. @ 180–200°F (82–93°C)	N/A
11 Flash (optional)	10 min. @ RT	10 min. @ RT	N/A
12. Final Bake Cycle	30 Min. @ 300–320°F (149–160°C)	30 min. @ 280–300°F (138–154°C)	25–30 min. @ 250–280°F (121–138°C)

NOTES: N/A = Not applicable. RT = Room Temperature
* = Number of coats may vary to achieve final film build
** = Initial or prebake cycle sometimes used after each application of color coat

finishing job. For the assembly plant, which has experience primarily with finishing metal parts, important differences exist when determining dry film build and oven baking temperatures. The dry film build is more critical on SMC parts than it is on metal parts. That is primarily due to the thermal characteristics of SMC. Optimum thickness builds for the dry films normally range between 1.8 to 2.5 mils for lacquer topcoats and 1.5 to 2.0 mils for enamels. Those have resulted in system performances and acceptable color matches in applications where steel components and FRP require identical appearances. FRP parts are insulators that require more time for the part to reach the desired temperature during baking cycles. Metal parts follow the oven temperature profile more closely, allowing a uniform release of paint solvents during the baking cycle. With FRP laminates there is a tendency for the paint film to "skin over" and trap solvent below it. That problem can be solved by selecting a paint system having a solvent composition that closely matches the thermal characteristics of the FRP laminate to achieve a more uniform release of the solvents. Initial baking cycles have been used successfully in preventing "skinning over" of paint films where solvent composition balance cannot be adjusted. Initial or prebake cycles are shorter in time and lower in temperature. They precede the final baking cycles of the finishing operation. Initial or prebake cycles are usually maintained at 180–200°F for 7–10 minutes. That procedure allows hardening of the paint film to the point where repair work, if necessary, can be done and establishing a controlled time and temperature cycle to improve solvent release. That procedure substantially minimizes such painting defects as solvent "boil" or "pop."

A relatively new topcoat procedure is to apply a basecoat/clearcoat (BC/CC), as is done on the Corvette and other models. All of the adhesion to the primer and the color is contained in the basecoat. A protective clearcoat is applied over the basecoat. The procedure, which results in additional gloss and better weatherability, will be discussed in more detail later.

Automotive Repair Paints

Some quantity of FRP parts will suffer handling damage or collect dirt during the priming operation. Such minor defects can be sanded smooth and spot primed with air drying primer that is formulated to match the color and gloss of the original bake primer. Small delaminations and blisters usually are pierced through the nonshow side of the part with a drill to allow volatiles to escape. A catalyzed epoxy then can be injected into the air space and the repair allowed to cure. Holes or cavities should be filled with a filled catalyzed epoxy repair material, such as Thermoset 100 and 101 (Thermoset Plastics Inc.) or the equivalent. Spot primers should not be used on the latter two repairs, but the repaired area should be leveled by sanding with 320 grit paper, and the entire part should be reprimed and rebaked in the normal production equipment. This will ensure that all the repair materials have been cured, and if any air remains that swells in the baking oven, the part should be scrapped. Repairing topcoated parts containing minor defects usually is done on the assembly line by removing the topcoat paint, sanding the affected area with 320–360 grit paper, tacking off the area, blowing deionized air on the spot, and recoating and rebaking the part. If the minor defect is discovered after the automobile has left the assembly line, spot repairs can usually be made by sanding the affected area smooth, tacking the area, spraying the same production paint over the repaired area, and baking the paint with infrared lamps. That procedure generally will not work with metallic lacquer topcoats, and the entire panel must be refinished.

FRP Substrate Paint Adhesion Improvement

The development of low shrink additives for FRP has resulted in smooth surfaces that require very little preparation before painting. They certainly have made it unnecessary to sand the surfaces before prime coats are applied, as was the original

custom. It has been observed that not all thermoplastic additives affected the surfaces to the same extent and their ability to accept paint varies. One laboratory investigation involved the following materials:
1. BMC-A Noncuring linear polyester + PVC
2. BMC-B High impact polystyrene + polyethylene
3. BMC-C Styrene-acrylate copolymer.

It was found that BMC-A accepted paint very well but had marginal molding properties by compression molding (porosity prone); BMC-B had better molding properties in that the incidence of porosity was reduced but had poor primer adhesion and enamel topcoat popping, and water blisters resulted. BMC-C had better molding properties and readily accepted paint. In an attempt to improve the primer adhesion of material BMC-B, it was found that the principal cause of primer adhesion, enamel popping, and water blistering was the low shrink additive used. Primer adhesion was improved by adding a polar monomer to the formulation, such as acrylamide or an ether of methylol melamine such as Cymel 303™ (American Cyanamide). The inclusion of rubber-type polymers in the FRP formulation in addition to providing "toughness" also improves primer coat adhesion to the substrate.

Application Equipment

Spraying has been the accepted method for large-scale application of paint. Spraying is the movement of paint from a nozzle to the part in the form of very small droplets. When the tiny droplets reach the part, they flow together to form a continuous wet layer. That layer, after it loses its solvent, is the paint film. In powder coatings tiny solid particles are sprayed onto the part, melted, and then solidified by cooling to form the final paint film. In some of the newer higher solids paints their viscosity must be reduced by heating as they are applied in order to spray them. Generally, airless systems are used for those materials. The latest high-volume pigmented paints can not be sprayed by conventional equipment and are projected toward the part by being impacted on very high spinning cones. Those applicators are referred to as "bells" or "cones."

A conventional air-spraying system consists of the air compressor, air supply lines, air regulator, paint pot, paint supply lines, and spray guns. With that system air is supplied to the gun and to the paint pot. At the gun this air mixes in a violent manner with the paint breaking it into small droplets. The air pressure also provides the energy to propel the droplets toward the part to be covered. A chiller of sufficient capacity to remove all water should be installed on the compressed air reservoir, and a filter must be installed on the air inlet lines ahead of the spray guns to ensure that only clean, dry, compressed air is furnished to the paint line. The air regulator allows the operator to select air at different pressures for the fluid and atomizing air lines. The paint pot usually is a pressurized container holding the paint. Air pressure forces the paint to the gun. Air line pressure, gun model number, tip number, fluid flow rate, and paint-thinned viscosity all should be specified on the Standard Operating Procedure or job card for each part to be painted.

With airless spray equipment there is no air line to the gun. Paint is supplied to the gun under very high fluid pressure, typically 1,000–4,000 psi. When the pressurized paint leaves the fluid tip, it enters the much lower air pressure region in front of the gun. That pressure change causes the paint to atomize.

Sometimes heated paint is used either on air or airless systems. The higher operating temperature lowers the paint viscosity and allows higher film build per gun pass. Heated paint systems usually operate in the 100–150°F (38–65°C) range.

With the increased use of higher solids paints and waterborne paints to meet environmental restrictions new application equipment became a necessity. Electrostatic atomizers have been used to dispense solvent systems, but the earlier units used electrical motors for rotation, and speeds were dependent on available motor windings. Smaller bells rotating at very high speeds have been developed to handle those coatings systems. Scharfenberger (Ransburg) and Cowley (Binks-Bullows) have published technical papers that describe the parameters of those new high-speed turbine units.

An important piece of support equipment is the spray booth. Generally it is of the down draft type with a water trap under the booth to catch waste paint. Air into the booth must be filtered and heated in winter months. The purpose of the booth is to keep waste paint from contaminating the remainder of the plant and dust from contaminating the paint line.

Paint Defects

Because of the possible confusion in terminology a definition of defect terms has been included here (see Table 9.8). Also included are several micrographs (see Figure 9.4) that illustrate some of the defects.

Defects That Appear During Painting

The formation of bubbles in the topcoat during the topcoat baking operation is the most common paint defect attributed to porosity in the FRP substrate. It is caused by solvent vapor pressure, which develops after the topcoat has set up to the point where it will no longer allow the escape of the vapors by diffusion. The reason that solvent is not given off during the flash period is that it is both adsorbed in pores (cracks, holes, or unwet glass fibers) in the FRP or adsorbed in agglomerates of thermoplastic low shrink additive, where it is retained until the heat of the baking oven causes it to boil. In some ways it resembles "solvent boil" in that both are caused by the formation of this solvent-impervious skin on the topcoat surface before all the solvent can leave. The difference is that solvent boil occurs only in the topcoat layer, is spontaneous, and can be remedied by adding retarders to delay the formation of the impervious film, while bubbles caused by adsorbed solvent are continuous and are not so easily remedied.

The shape rather than the size of the pore determines if it will cause bubbles. In a controlled experiment using a mold

TABLE 9.8 Paint Defect Definitions and Descriptions

Porosity: Any visible blemish on a painted surface caused by a physical defect on or near the surface of the FRP substrate. This usually appears as a depression that has the same dimensions as the hole, crack, roughness, or dry glass on the unpainted surface. Such a condition is shown in Figure 9.4 in both a micrograph of the painted surface and cross section of the part.

Paint Bubble: This is a raised area of paint filled with gas and appears during the baking of the topcoat. In appearance it resembles the solvent boil but is usually isolated, whereas boils occur in clusters. When the pressure exceeds the strength of the film, these bubbles break to form cavities, which are called "pops." They have a craterlike appearance, often with primer showing in the bottom of the hole.

Solvent Boil: This is a form of bubble caused by improper application of the topcoat because of poor selection of thinners, too short a flash time, or too thick a film. It is easily identified, since it usually appears in clusters in areas where paint is obviously too thick. Solvent boils are shown in Figure 9.4 as a micrograph of a cross section of the paint film.

Dirt: Contamination from the air in the paint booth or baking ovens results in a deposit of dust particles, rust, etc., in the wet enamel finish to form blemishes that can resemble bubbles and holes caused by prosity, like those shown in Figure 9.4. This generally is the result of poor house keeping in the assembly power paint department.

Delamination: Delamination is a bubble in the interior of the FRP part and is caused by air that was trapped as the charge flowed toward the extremities of the mold. This defect appears during the paint baking operation if the temperature exceeds 300°F.

Blisters: These resemble large bubbles except that they are caused by pressure from watery liquid instead of gases. Like bubbles, they can break and form holes or collapse when the water evaporates to form craters.

having holes of different shapes and sizes, it was found that those having a cylindrical shape and tapering to a point at the bottom will form bubbles. Larger and shallower holes of other shapes did not.

The chemical composition of the thermoplastic additive determines if it will absorb and retain solvent long enough to cause bubbling. It was found that polar thermoplastic materials, such

FRP Painting 241

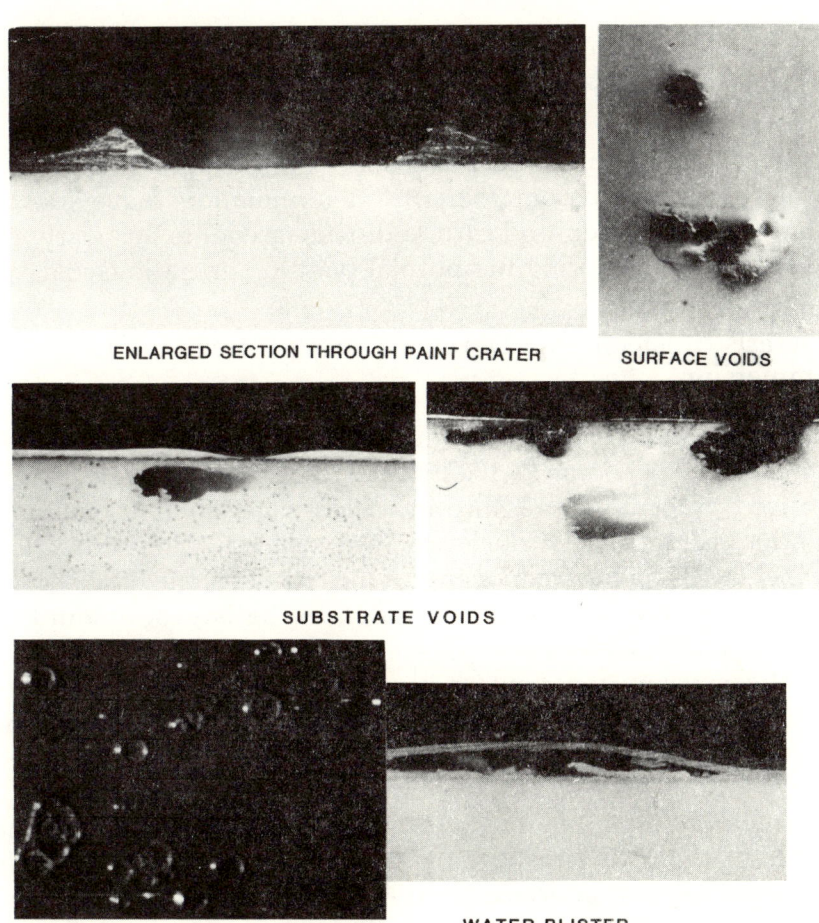

Figure 9.4 Paint Defects

as polymethyl methacrylate and vinyl chloride copolymers, released solvents much faster than did nonpolar thermoplastics, such as polystyrene and polyehtylene, and also caused less bubble formation.

Gases absorbed in the FRP molded part do not cause topcoat defects. They were found to be given off during the preheating treatment before the topcoating operation. Also, no relation between the amount of gases absorbed in the part with defects was found. Gas chromatography determinations have shown that styrene, water, and carbon dioxide accounts for nearly all gases given off during the normal painting operations.

Defects (Blisters) That Develop During Water Immersion

The formation of tiny blisters on the topcoat during water immersion (240 hours @ 100°F [38°C] in distilled water) is mainly caused by osmotic pressure in which water-soluble materials under the topcoat surface are responsible. Water extraction experiments on the components of the FRP substrate have shown that the polyester resin contributes to the largest amount of water-soluble materials.

The mechanism of water blister formation starts with the absorption of water into the topcoat, which in the case of acrylic type paint takes 2 to 3 days and the amount of water absorbed is about 0.3–0.4%. The swollen topcoat then acts as a semipermeable membrane through which water can pass from the bath to the FRP substrate-paint interface, but water-soluble materials at that interface are not allowed to pass through the film into the bath. Osmotic pressure develops, and the paint film is separated from the substrate and rises to form a blister. The size of the blister is determined by the amount of water-soluble material and the osmotic pressure of water in the bath, the thickness and modulus of the paint film, and the adhesion of the paint to the substrate. The blister continues to grow until the combination of adhesion and modulus forces balance between those of the osmotic pressure.

Some sealers and primers are more resistant to blister formation than others because they have a tighter adhesion to the substrate. They are most effective when applied directly to the FRP substrate instead of on top of the primer.

Blister formation can also be reduced by including certain additives to the FRP formulation. Those can be urea- and melamine-formaldehyde condensates, such as Cymel™ 303 (American Cyanamide) or nitrogen-containing monomers, such as vinylpyridine, dimethyl aminoethyl methacrylate, or acrylamide. That is due to increased adhesion of the primer to the FRP substrate.

A nondestructive inspection test on the unpainted part in order to locate defects that need repairing can be made. The test involves coating the unpainted part with a water solution of fluorescent material and removing that part which remains on the surface but not that part which is deposited in the crevices. Water was chosen as the solvent since it is not absorbed by agglomerates of thermoplastic additive on the surface. Under ultraviolet light, crevices and holes are easily seen, yet in visible light, the appearance of the part has not changed.

Case Histories Before Low Profile Systems

National Cash Register

The first SPI paper on painting FRP parts was delivered by Nelson and Muntion of NCR in 1953. Table 9.9 details the process conditions used. The extralong bake cycle was required since the finishing operation, in addition to handling aluminum machine housings containing the FRP panels, also was used to finish wood parts. Those data are presented for historical purposes only.

Outboard Marine Motor Shrouds

The Gale Products Division of Outboard Marine Corporation, Galesburg, Illinois, built a facility for producing preform reinforced polyester outboard motor shrouds. The plant had a

TABLE 9.9 NCR FRP Wrinkle Finishing Schedule

a. Clean part: use soft cloth and wet with naphtha or toluene.
b. Dry part: at least 5 minutes @ room temperature.
c. Fill pinholes: apply Lowe Bros Paint Co. wood filler #31711 reduced to working consistency with naphtha. Allow to dry until tacky; pad into pinholes and remove excess filler.
d Dry: 2 hours @ 150–160°F (66–71°C)
e. Prime coat: apply a medium wet coat of Lowe Bros. Thermo-Sprafil C-15780 reduced with lacquer thinner to a viscosity of 70–75 seconds in a #7-D cup.
f. Bake: at least 45 minutes at 150–160°F (66–71°C)
g. Sand: using 220 grit dry paper, sand only places having deep ridges, dirt, or other defects.
h. Enamel coat: apply a double coat of Hanna Industrial Finishes Co. gray 31 rough enamel all over.
i. Bake: 20 hours @ 150–160°F (66–71°C).

capacity of 1,800 motor shrouds per working day and began production for the 1959 model year. Half the shrouds were sent to the Johnson Motor Division in Waukegan, and the remainder were sent to the Evinrude Motor Division in Milwaukee. At the assembly plants the shrouds were cleaned in a trichloroethylene (TCE) vapor degreaser operating at 188°F (91°C) and then received a 3–4 mil dry film epoxy surfacer-primer, which was applied with a hot spray gun at 150–160°F (66–71°C) and then baked for 30 minutes at 240°F (116°C). The entire shroud surface was wet sanded with an orbital sander to remove orange peel from the primer coat. A catalyzed epoxy topcoat was then applied at 160°F (71°C) and baked for a minimum of 45 minutes at 230–240°F (116°C). The chief problem was the rough surface appearance of the shroud, as compared with the previously used aluminum die casting, despite the use of surfacing veil over the preform. The rough surface with its fiber pattern was the reason 3 to 4 mils of primer were applied and then thoroughly wet sanded to provide a smoother appearance. Occasionally a shroud fell off the paint line at the cleaning station and had to be scrapped. Several hours' immersion in hot TCE turned the rigid tough shroud into a rubbery configuration.

TABLE 9.10 Comet Hood Final Finishing Schedule

a.	Apply 3 coats wet-on-wet ESB-M6J17A (Ford specification) alkyd primer reduced to 16–20 seconds Ford D cup
b.	Apply 1 coat (wet-on-wet) ESB-M99J222A (Ford specification) enamel sealer
c.	Monobake: 40 minutes @ 290°F (143°C)
d.	Wet sand all over
e.	Dry hood
f.	Inspect
g.	Package for shipment

Comet Hood Program

One FRP molder produced top color-coated fiberglass hoods for the 1966 model year Mercury Comet. As production began, it was soon realized that color topcoats could not be applied at the molder's plant and have this color match the color and gloss on a steel car, even though, for some special test parts, the same batch of paint was used for both the hood and the balance of the car. Topcoating was discontinued on those hoods in the molder's plant, and for the balance of the model year they were shipped only with a prime coat, and a sealer-inspection coat to permit better surface inspection before shipment. The final painting schedule is shown in Table 9.10.

Table 9.11 shows a typical analysis of defects for a weekly period during the early part of the program. A task force studied the inherent production problems with the hoods, and although some improvements were made in the molded part surface finish, the problems were not completely solved and this program turned out to be a financial loss for the molder.

Miscellaneous

Several other SPI papers (Savage, Irwin, & Austin[91] and Savage & Weiner[102]) are listed in the annotated bibliography. The chief problem with those early painting efforts was the requirement for a very heavy primer built to help obscure fiber pattern, and a great deal of hand sanding was required to level the primer, which made painting FRP parts uneconomical as compared with similar metal parts. Porosity was the principle defect observed in the FRP substrate. A considerable number of reports were published on painting the Corvette and Avanti bodies. The problems were the same as reported above, and no point will be served in needless repetition.

Case Histories With Low Profile Systems

Ford Fender Extension Program

The Marion, Indiana, plant of General Tire & Rubber Company produced Ford fender extensions for the 1968 model year. There were 24 individual molds involving 6 different fender extensions. From the summer of 1968 through February 1969 an average of 1,250,000 pounds (5,682 metric tons) of low profile BMC per month were compression molded. The fender extensions were deflashed at the press, each received an average of seven studs, and they were then hung on the paint line and wiped with a soft cloth and isopropanol. A solution of Ransprep™ 100 in isopropanol was hand applied to the extensions to make them electrically conductive (today a furnace black pigment dispersed in a linear polyester grinding vehicle would be added to the BMC formulation). Figure 9.5 consists of a block diagram and flow sheet of that finishing operation. Surface smoothness

TABLE 9.11 Comet Hood Major Defects

a.	Pin Holes	33.3
b.	Contamination	17.9
c.	Solvent Pops	17.6
d.	Sanding Scratches	8.3
e.	Paint Runs	7.6
f.	Thin Paint	5.7
g.	Improper Repair	2.1
h.	Resin Cracks	1.4
i	Other	6.1
j.	Total	100

FRP Painting 247

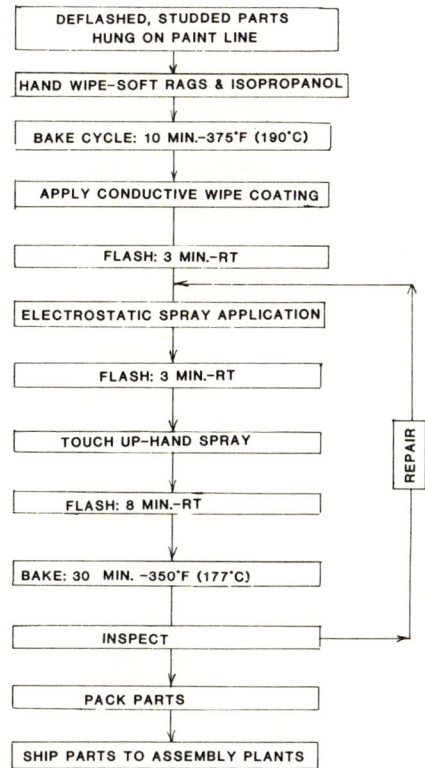

Figure 9.5 Priming Procedure: Ford Fender Extensions

of the parts as molded, physical properties, esthetics of the part on the automobile were all satisfactory. The only problem was porosity in the parts as molded. Too many parts had to be repaired. Ford would not allow an air dry touch-up paint over a repair, and the part had to be recirculated through the paint line after repair. That procedure ran up the manufacturing cost. When Ford was advised that the piece price would be increased for the 1970 model year to offset the high repair costs, it converted the fender extensions to die cast zinc.

Pontiac Front End Problem-1971

During the 1971 model year, the Pontiac Motor Division of General Motors had SMC front ends on two car models that

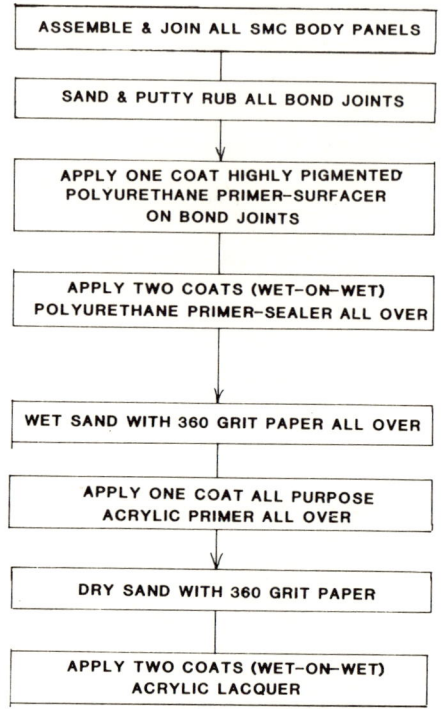

SOURCE: DISPLAY CAR AT 34TH SPI RP/CI EXHIBIT

Figure 9.6 Corvette Finishing Schedule, 1979

were assembled and finished at three different major assembly plants. Those front ends were made by six different molders and were prime coated by the molder before shipment to the assembly plant. A severe topcoating problem developed at all three assembly plants, although the severity was different at each plant. The worst problem was at the Pontiac Division assembly plant in Pontiac, Michigan. To obtain all the facts firsthand, the molders, together with a few major raw materials suppliers, formed a task force that monitored the process at two locations, two shifts per day, for a 3-week period of the Pontiac, Michigan, plant. All parts were inspected as they were hung on the paint line and after they emerged from the final baking oven. The part was divided into 7 sections,

FRP Painting 249

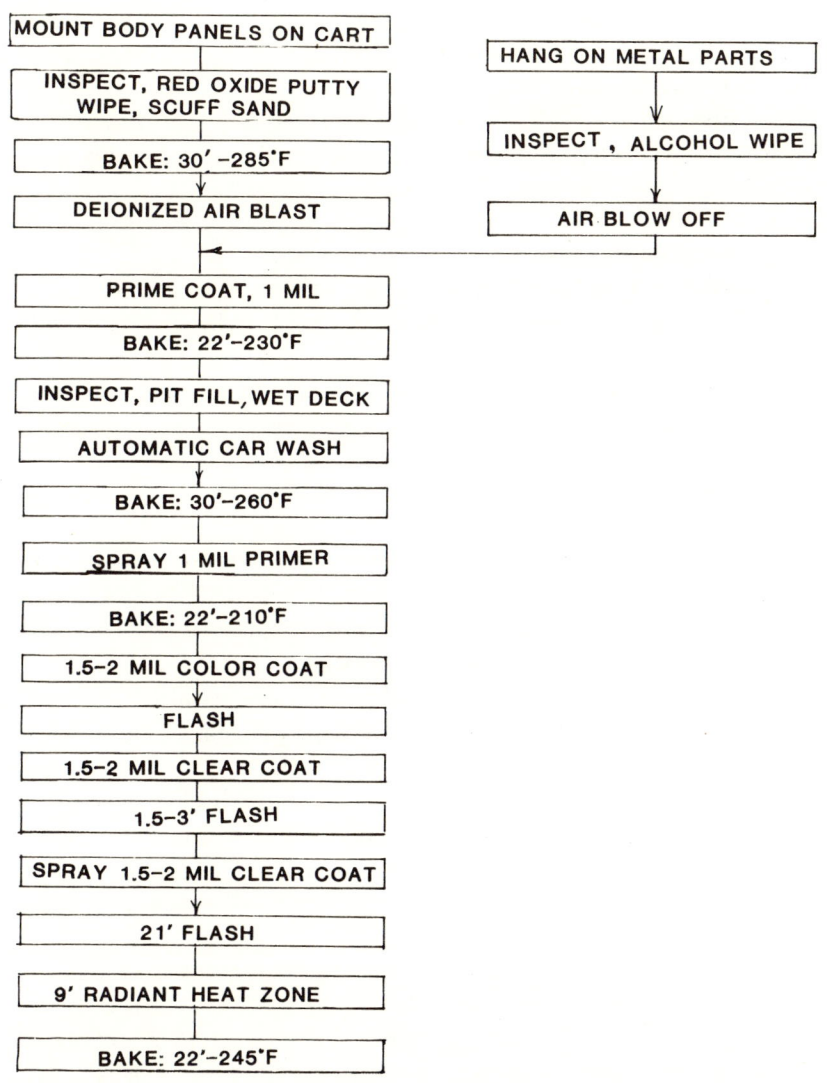

Source: Schrantz, INDUSTRIAL FINISHING ,(Mar., 1984)

Figure 9.7 Corvette Current Body Paint Schedule

and all defects were recorded by part area, by molder, by top coat color code, and by part serial number (added at the first inspection station). At the conclusion of the information collection stage the data were analyzed and the following were found:

1. There was a difference in the molder part quality level (2 molders produced most of the major defects).
2. Top coat "solvent boil" was worse for darker colors than for lighter colors.
3. Paint defects were more prone to occur in vertical part areas as molded (less pressure).
4. Porosity was a problem for all SMC molders.
5. Painting process conditions needed to be studied further.

As a result of that task force work the two worst molders ceased to be suppliers of SMC parts to the automotive industry. One molder even sold his SMC impregnator. Each of the remaining molders was asked to review his process and procedures in an effort to improve his parts. General Motors established an "in house" SMC committee to study SMC problems. General Motors and Owens-Corning Fiberglas Corporation, with the assitance of the custom molders, reviewed the painting procedures of other GM assembly plants and made a designed experiment paint study at the OCF Granville Technical Center. The results of this study revealed:

1. Current low profile FRP parts can be and are being painted together with steel parts under proper conditions.
2. The solvent thinner system used with acrylic reflow lacquer top coat is critical with respect to FRP. Best results were obtained with: a. 75/25 duPont 37988/butyl cellosolve, b. Gage Solvent Co.'s GP-36308, c. 75/25 duPont 37988/butyl cellosolve acetate.
3. An initial low-temperature bake (10 minutes @ 185°F [85°C] hot air or infrared) before reflow was found to be essential.
4. Paint defects occurred more frequently on vertical painted surfaces than on horizontal painted surfaces.

OCF Painting Study Results

After the Pontiac SMC paint problem had been resolved, the OCF Technical Center laboratories in Granville engaged in an exhaustive FRP paint program of several years' duration. At its conclusion OCF put together a slide film presentation for the FRP industry to announce its findings as follows:

1. The difference in thermal conductivity of FRP and steel requires that they be handled differently on a paint line. Steel parts will heat up more quickly under the same conditions, but FRP will maintain a more even temperature.
2. Prime coats on FRP should be no thicker than 1 mil.
3. Prime coats should be baked a minimum of 10 minutes at 10–15°F (6–10°C) higher than the final temperature the part will see during topcoat baking.
4. Thermosetting prime coats are much preferred to thermoplastic prime coats, since they have better solvent resistance, better heat stability, and higher properties.
5. Permissible topcoat color film thicknesses:
    FRP   1.8–2.8 mils
    body steel   1.8–4.0 mils
6. Special attention must be paid to solvents in topcoats so that they can be completely evaporated during the oven cycle and not be entrapped by skinning effects.

Miscellaneous

No major painting problems have existed at any automotive assembly plant now for several years. The earlier problem areas have been listed as a matter of history, and the FRP and automotive industries in the United States have learned from those problems. One of the keys in the learning process was the willingness of the custom molder, the raw materials suppliers, and the automotive assembly plants to share knowledge and help when a problem existed. Without that cooperative spirit the industry would not have solved the many problems it has been exposed to during the past two decades. The fact that those problems have been solved is attested to by the automotive companies themselves. Today there are several captive SMC plants operated by automotive divisions.

Annotated Bibliography

Introduction

The items in this bibliography are the result of an intensive literature search and are arranged chronologically. Only the first two initials of the author's names are listed to save space. The titles of all books and journals are italicized.

Abbreviations Used In This Bibliography

Anon.	anonymous
Abs.	abstract
BPF	British Plastics Federation
BMC	bulk molding compound
CAB	cellulose acetate butyrate
DMC	dough moulding compound (European equivalent of BMC)
FRP	fiberglass reinforced plastics
LP	low profile
LS	low shrink
LSW	low shrink wet lay-up system
pp.	pages
RETEC	Regional Technical Conference of SPE
RP	reinforced plastics
RTPMC	Reinforced Thermoset Press Molding Committee of SPI

253

SAE   Society of Automotive Engineers
SMC   sheet molding compound
SPE   Society of Plastics Engineers
SPEJ   Society of Plastics Engineers Journal
SPI   Society for the Plastics Industry

1. Bigelow, M. H., "Alkyd Hot Molding Compounds," *Mod. Plas.*, Oct. 1948, pp. 85–87.
    Abs. First article found on Plaskon alkyd molding compounds.
2. DeVore, H. W., "Alkyd Molding Compound Goes to Work," *Mod. Plas.* (27-3), Nov. 1949, pp. 81–83.
    Abs. Plaskon's alkyd molding compound is described, and its use in electrical parts is documented.

1950

3. Edmunds, W. H., "Plastics for Switchgears," *SPEJ*, Apr. 1950.
    Abs. This paper discusses the use of FRP in switchgear apparatus.
4. White, R. B., "The New Horisons for Fiberglass and Polyesters in the Electrical Field," SPI RPD 5th Ann. Conf., Cleveland, Ohio, Jan. 10–12, 1950, Sec. 11.
    Abs. First mention of fiberglass molded parts (premix).

1951

5. Parkyn, B., "Polyester Resins in the Electrical Industry," *Brit. Plas.*, Feb. 1951, pp. 47–49.
    Abs. One of the first European articles to mention DMCs.
6. Weaver, W. I., U.S. Patent 2,549,732, "Production of Polymerized Unsaturated Resin Materials of Superior Water Resistance," Apr. 17, 1951.

Abs. Pigmented unsaturated polyester resins thickened with Group II metal oxides, hydroxides, or carbonates used as castings or adhesives.

7. Frilette, V., U.S. Patent 2,568,331, "Copolymerization of Styrene and Unsaturated Resins," Sept. 18, 1951.

    Abs. The use of Group IIA oxides or hydroxides to harden unsaturated polyester resins without the use of external heat is described.

1952

8. Bigelow, M. H., and Nowicki, P. E., "A New Reinforced Alkyd Plastic," SPI RPD 7th Ann. Conf., Section 6-I (3 pp.), Apr. 9–11, 1952.

    Abs. Alkyd 440 molding compound is described. Physical and electrical properties are quoted.

9. Slayter, G., "A Look at the Future of RP," Ibid., Sec. 23.

    Abs. Production parts weighing 75 pounds have been molded of premix compounds.

1953

10. Fisk, C., U.S. Patent 2,628,209, "Process for Increasing Viscosity of Uncured Alkyd Copolymer Resinous Mixtures and Products," Feb. 10, 1953.

    Abs. The use of Group IIA metal oxides to harden polyesters without the use of external heat is described.

11. Dietz, A. G., "Glass Fiber Polyester Molding Materials," Proceedings SPI RPD 8th Ann. Conf., Feb. 18–20, 1953.

    Abs. Formulations and physical properties are given for four compounds.

12. Nelson, B. W., and Muntion, C. B., "Finishes for and Surface Qualities of Fiber Glass Reinforced Moldings," Proceedings 8th SPI RPD Meeting, Chicago, IL, Feb. 18–20, 1953, Sec. 5.

13. Borro, E., "Closed Mold Molding Techniques," *SPEJ* (8 pp., 3 photos, 11 diagrams), Apr. 1953.

    Abs. This article stresses proper part and mold design with emphasis on gate and runner design for multiple cavity parts.

1954

14. Nelson, B. W., "Available Finishing Systems for RP," Proceedings 9th SPI RPD Meeting, Chicago, IL, Feb. 3–5, 1954, Sec. 5.
15. White, R. B., "Glass Premix Molding," SPI RPD 9th Ann. Conf., Chicago, IL, Feb. 3–5, 1954, Sec. 11-D (8 pp., 11 photos, 3 diagrams).

    Abs. This paper discusses some of the applications for premix compounds and why these materials are being used in these applications.
16. Reiling, V. C., "Transfer Molding of Premix Molding Compounds," Ibid., Sec. 11-E (3 pp.).

    Abs. The techniques of transfer molding premix compounds are discussed.
17. Sheppard, H. P., "Polyester Glass Molding Compounds," *Mod. Plas.*, Apr. 1954, pp. 121–135.

    Abs. Premix compounds and raw materials are described. Statistical charts of physical properties of various glass contents are provided. The coefficient of variation increases as glass content increases.
18. Erickson, W. O., and Ahrberg, W. R., *Mod. Plas.*, Nov. 1954, pp. 125–133.
19. Morgan, P., *Glass Reinforced Plastics*, Iliffe & Sons, London, 1954 (248 pp.).

    Abs. A brief history of DMC development, molding practices, and molding problems is outlined.
20. Sonneborn, R. H., *Fiberglass Reinforced Plastics*, Reinhold Pub. Corp., New York, 1954, pp. 62–63.

Abs. A very short description of the premix process is given together with a list of advantages and disadvantages.

1955

21. Culweir, E. F., "Properties and Use of RP and Other Related Materials," Proceedings SPE 11th NATEC, Atlantic City, NJ, Jan. 1955, pp. 201–213.

    Abs. A description of premix materials is given, and a table is provided showing their physical and electrical properties.

22. White, R. B., "Glass-Premix Compounds for Molded Parts," *Elec. Mfg.*, Mar. 1955 (8 pp.).

    Abs. Premix compounds consisting of glass fibers, resins, and filler produce molded parts both for insulating and for structural uses with significant property advantages in many applications. Typical applications are illustrated.

23. Anon., "An Introduction to Premix Compounds," Barrett Div., Allied Chem. & Dye Corp., New York.

    Abs. This bulletin describes the art of premix molding, pointing out its limitations and advantages and describing the apparatus and techniques involved.

24. Erickson, W. O., and Ahrberg, W. R., "Reinforced Polyester Premixes—How to Mold Them," *Mod. Plas.*, Nov. 1955, pp. 125–131/255.

    Abs. Partially based on Item 23 above. A good illustrated discussion of preparing premix compounds is given.

25. DuBois, J. H., "Use of the Right Material," *SPEJ*, Oct. 1955, p. 26.

    Abs. This paper discusses the advantages of plastics materials, such as weatherability, thermal insulation, electrical insulation, colorability, etc., and stresses the use of the proper material to meet the application.

26. Shannon, R. F., and Biefield, L. P., "Highly Filled Fibrous Glass-Reinforced Polyester Molding Compound," *Modern Plastics*, Nov. 1955, pp. 133–138/258 (5 pp).

Abs. A series of molding compounds containing 5% glass fibers and various inert fillers is discussed. More highly filled glass compounds made in a Brenner chopper and in a dough type mixer also are included.

1956

27. Hansen, A. M., "Large Scale Application of Premix," SPI RPD Proceedings 11th Ann. Conf., Atlantic City, NJ, Feb. 5–9, 1956, Sec. 13-B.

    Abs. The use of sisal reinforced premix for automotive heater housings, estate wagon garnish moldings, and battery hold down rings is discussed.

28. Pelham, O. W., "Some Notes on the Properties of Premix Compounds," Ibid., Sec. 13-C.

    Abs. Woodall reports using 60,000 pounds per day of premix for small automotive interior parts. The nonuniformity of physical properties in a molded flat sheet is examined.

29. Erickson, W. O., and Ahrberg, W. R., "An Introduction to Polyester Premix Molding," Ibid., Sec. 13-D (6 pp., 9 photos).

    Abs. Apparatus, ingredients, and techniques involved in premix molding are discussed. Two formulations are provided.

30. Anon., "Premix Molding Compounds," Tech. Bull. L-6-56, Resinous Prod. Div., Rohm & Haas Co., Philadelphia, PA (5 pp.).

    Abs. This bulletin describes premix compounds, the raw materials required, preparation suggestions, moding notes, and 3 formulations.

31. Greig, J. W., "Matched Metal Die Molding of Premixed Reinforced Molding Compounds," *SAE Transactions* (1956).

1957

32. Doyne, R. F., "Premix Molding Techniques and Tooling," SPI RPD 12th Ann. Conf., Feb. 5–7, 1957, Sec. 17-A.

    Abs. Molding pressures of 259–1,000 psi, flash clearances of 3–7 mils, venting of dead-end areas of flow, and compression molding rather than transfer molding of parts are recommended.

33. Crenshaw, J. B., "Formulating Polyester Molding Compounds," Ibid., Sec. 17-B.

    Abs. The object of this paper is to present methods and materials for making suitable polyester premix compounds economically. A total of 15 formulations is given.

34. Erickson, W. O., and Ahrberg, W. R., "A Study of Various Reinforcing Materials in Polyester Premix Compounds," Ibid., SEc. 17-C.

    Abs. Data are presented on the effects of molded samples vs. machined samples on nylon, Dacron, sisal, asbestos, and glass reinforcements; on glass fiber lengths and weld lines on the physical properties of premix compounds.

35. Harvey, J. L., "The Place of Premix in the Molding Industry: Review of Important Applications," Ibid., Sec. 17-E.

    Abs. This paper presents the 1957 uses of premix compounds for automotive, electrical, and tool handle applications. Profusely illustrated.

36. Sadler, K. B., "Co-operative Development of Improved Test Methods for Premix," Ibid., Sec. 17-D.

    Abs. The design of a new test bar and mold in which to make this bar is discussed.

37. Werkheiser, R. L., "Comparative Mechanical Properties of Polyester and Phenolic Based Premix Compounds," Ibid., Sec. 17-F.

    Abs. A total of 13 premix formulations involving both polyester and phenolic resins is presented together with physical properties.

38. Guzetti, A. J., "Shrinkage of Thermosets," *Mod. Plas.*, Feb. 1957, pp. 111–113, 116–130.

    Abs. The effects of molding variables, part design, charge placement, etc., are discussed with emphasis on those effects on shrinkage.

39. Strasser, F., "Design Tips for Plastic Moldings," *Machine Design*, Apr. 18, 1957, pp. 144–149.

    Abs. Plastic part design tips are presented including side wall taper, wall thickness, molded-in holes, strengthening ribs and fillets, parting line location, inserts, and tolerances. Illustrations of "good" and "bad" practices are included.

40. Colao, J. J., "Premix Your Own Phenolics," *Mod. Plas.*, Nov. 1957 (5 pp., 10 photos).

    Abs. This paper shows how to compound phenolic premixes to replace molding powders at considerable cost savings. Recipes, equipment, compounding temperatures, cycles, and applications are discussed.

1958

41. Jackson, R. S., "Reliable Properties Can Be Guaranteed Despite Many Cantankerous Variables," Proceedings SPI RPD 13th Ann. Conf., Chicago, IL, Feb. 4–6, 1958, Sec. 3-E (8 pp.).

    Abs. This paper discusses some of the statistical quality control techniques that can be used to ensure the quality of FRP products.

42. Wittman, L., "Report on the Premix Standards Committee," Ibid., Sec. 9-B (2 pp.).

    Abs. This paper covers the evauation program by the Premix Committee to develop a new method of preparing molded specimen for impact and flexural tests. No data are reported.

43. Calderwood, R. H., and Sheppard, H. R., "Physical Properties of Transfer and Compression Molded Premix Materials," Ibid., Sec. 9-C (10 pp., 5 photos).

Abs. Test bars were prepared from premix compounds by both transfer and compression molding techniques. Glass fiber content of the premixes were varied. In general, it was found that transfer molding resulted in higher flexural strengths and lower impact strengths, and both could be attributed to fiber orientation.

44. Proudfit, C. W., "Applications and Suggested Uses of Polyester Premixes," Ibid., Sec. 9-D (4 pp.).

    Abs. The premix market for 1956–1958 is estimated. Some applications are presented, and others are suggested.

45. Anon., "Selectron Premix Compounds: Two-Component and Three-Component Reinforcement Systems," PPG Industries, Pittsburgh, PA, Technical Report (26 pp.).

    Abs. The computer results of a designed experiment using glass fibers, asbestos, and sisal as reinforcements are presented. The effect of various combinations of these reinforcements on physical properties, handleability, appearance, and shelf life are tabulated.

46. Davies, J. D., Scott, K. A., and Sutcliffe, M. R., "The Manufacture, Properties, and Applications of DMC," British Plastics Fed. Meeting, London, Paper Q, Oct. 21–24, 1958.

    Abs. The raw materials used in DMCs are discussed. The advantages and disadvantages are listed, and manufacturing notes and properties are given. A good schematic presentation and physical property results of charge placement in a flat sheet mold are presented.

47. Torres, A. F., and Feuer, S. S., "A Method for Measuring and Predicting Cure Time of Premix Molding Compounds," Proceedings SPI RPD 14th Ann. Conf., Chicago, IL, Feb. 3–5, 1959, Sec. 9-A (4 pp.).

    Abs. An adaptation of the SPI cure characteristics test method for polyester resins is described for use on premix compounds.

48. Duprez, H. J., "Survey of Mold Design for Premix," Ibid., Sec. 9-B (6 pp., 8 sketches).

Abs. This paper describes the basic approaches and design trends used in fabrication of molds for premix compounds.

49. White, R. B., "Problems with Premix Moldings," Ibid., Sec. 9-C (4 pp.).

    Abs. Glastic outlines a program to sell premix compounds to other molders together with an engineering service.

50. Anon., "New Mass-Production Plant for Reinforced Plastics Molding," *Modern Plastics*, Feb. 1959, pp. 87–92.

    Abs. Many innovations in plant layout, materials control, testing, molding, and finishing are features of a new Outboard Marine facility geared to produce 1,800 motor covers per day.

51. White, R. B., and Jackson, R. S., "Problems with Premix Moldings: Part I: Strength Variations and Voids," *Mod. Plas.*, Mar. 1959, p. 117 (9 pp.).

    Abs. This paper illustrates the strength variations resulting from fiber orientation, the tendency to include more voids in a molding as glass content increases, and the use of zebra charges to observe the flow of the premix.

52. White, R. B., and Jackson, R. S., "Problems with Premix Moldings: Part II: Cracks and What to Do About Them," *Mod. Plas.*, May 1959, p. 115, (5 pp.).

    Abs. Weld lines and flow line cracks are well illustrated. Suggestions are offered to overcome some of these problems by proper part design or charge placement.

53. White, R. B., and Jackson, R. S., "Problems With Premix Molding: Part 3: Sticking in The Mold, Warping, Inserts and Thin Wall Problems," *Mod. Plas.*, Jul. 1959, p. 110.

    Abs. Unsuspecting undercuts, poor mold finish, and no draft walls all are responsible for parts sticking in the mold. Sloppy operating knockout pins and plates frequently cause cocking of a part and difficult ejection. Sticking also results from a sudden mold temperature drop or when moisture is absorbed by the compound. Nonuniform glass distribution causes localized part warping and angular corners shrink and come out of the molds sharper than the angle built into the mold.

54. White, R. B., and Jackson, R. J., "Problems with Premix Molding: Part 4—Poor Fill, Deflashing, Special Problems, Small Parts and Mold Maintenance," *Mod. Plas.*, Sept. 1959, p. 113, (6 pp.).

    Abs. Voids occuring in blind pockets frequently can be eliminated by venting the mold with pins, etc. Proper part design greatly assists flash removal.

55. Cutler, N. A., "Premixed Polyester Molding Compounds in Europe," *British Plas.*, Aug. 1959, pp. 373–377.

    Abs. Raw materials, formulations, mixing procedures, and European applications are shown.

56. Jackson, R. S., "Electrical Premixes Come of Age," Proceedings 2nd Nat. Conf. Elec. Ins., Washington, D.C., Dec. 8–11, 1959, AIEE paper No. CP-5040, pp. 36–40.

    Abs. This paper points out that high filler, low glass premixes give the optimum combination of properties for electrical applications.

57. Anon., "Polylite 8183 Speeded Production for Compression Molders," *By Gum*, Reichhold Chem. Inc., White Plains, NY, Dec. 1959, pp. 18–20.

    Abs. Heater housing ducts made by Barnum Bros., Detroit, are shown.

1960

58. Cody, W. P., U.S. Patent 2,922,769, Jan. 16, 1960.

    Abs. Dimer acid-glycol polyester made thixotropic by the addition of a calcium salt.

59. Jackson, R. S., "Electrical Premixes Come of Age," Proceedings SPI RPD 15th Ann. Conf., Chicago, IL, Feb. 2–4, 1960, Sec. 16-E (9 pp., 13 photos).

    Abs. Similiar to Item 56 above. High filler, low glass content premixes are discussed.

1961

60. Doob, H., and Phillips, T. E., "Comparison of Reinforcements and Study of Certain Processing Variables in Premix Molding," Proceedings SPI RPD 16th Ann. Conf., Chicago, IL, Feb. 7–9, 1961, Sec. 9-A (10 pp.).

    Abs. HSI type glass reinforcement was found to give faster wet-out, greater loadability, and stronger knit lines than conventional hard chrome-sized fibers in premix compounds. Glass level, moisture content, mixing temperature, and mixing times were investigated.

61. Meyer, R. W., and Orkin, R., "Quality Control of the Mixing Operation at the Glastic Corporation," Ibid., Sec. 9-B.

    Abs. The use of reliable recording instruments at all important process stations to provide batch records is discussed.

62. Marszewski, C. A., and Leeper, S. E., "Polyester Premix: Performance in Relation to Resin and Formulation Practice," Ibid., Sec. 9-C (3 pp.).

    Abs. Techniques for minimizing fiber degradation and improving fiber wetting are outlined. Higher flexural and impact strengths were obtained by hot mixing the premix compound.

63. Vanderbilt, B. M., and Clayton, R. E., "Premixes Based on Hydrocarbon Resins," Ibid., Sec. 9-D (16 pp.).

    Abs. A liquid butadiene type polymer in vinyl toluene is described for premix compounding.

64. Shanta, P. L., "Flow of Premix Compounds," Ibid., Sec. 9-E.

    Abs. The adaptation of the ASTM flow cup test for premix compounds is described.

65. Pratt, B. D., "Evaluating Experimental Premixes," Ibid., Sec. 9-F (6 pp.).

    Abs. This paper describes a systematic approach to premix development using a punch card system to tabulate physical and electrical property data. The differences between molded and machined coupons are outlined.

66. Meyer, R. W., "Time Records of Mixing Operation Keep Check on Quality of Glass Polyester Molding Compounds," *Plas. Des. & Proc.* (9-1), Apr. 1961, pp. 26–27.

   Abs. This paper stresses the importance of time records of the BMC mixing operation as a permanent part of the QC records.

67. Marks, J. A., "Molded-Polyester-Glass Permits New Insulation Performance Standards in Switchgear And Controllers," *Insulation*, Aug. 1961.

   Abs. The use of premix to replace porcelain in electrical insulators is shown.

68. Anon., "Premix Compounding," Bulletin 50-3, Interchemical Corp., Cincinnati, OH, Dec. 1961 (13 pp.).

   Abs. Raw materials, mixing equipment and procedures, and formulations are discussed. This is one of the better such bulletins produced by resin vendors.

69. Anon., *Cellobond Polyester Resins*, Tech. Manual, British Resin Products Ltd., London, 1961 (22 pp.).

   Abs. A brief description of raw materials, mixing, molding, and applications of DMC are given.

1962

70. Seidel, L. H., "Defects in Glass-Polyester Compound Moldings—and How to Cure Them," *Plas. Des. & Proc.*, Jan. 1962 (2-1), pp. 12–16.

   Abs. This paper discusses some of the causes for defects such as voids, short shots, cracking, etc. A molding troubleshooting guide is included.

71. White, R. B., "Premix Compound Applications," Proceedings SPI RPD 17th Ann. Conf., Chicago, IL, Feb. 6–8, 1962, Sec. 5-A, (12 pp., 17 photos).

   Abs. Several case histories are presented of premix applications.

72. Thomas, J. J., and Crouch, R. T., "Asbestos in Polyester Premix Molding Compounds," Ibid., Sec. 5-B, (4 pp.).

Abs. Data are presented to show the value of asbestos fibers in premix compounds as the sole reinforcement and in combination with sisal and glass fibers.

73. Johnson, R. B., "Polyester High Strength Glass Premix-Variables Involved and Lines of Improvement," Ibid., Sec. 5-C.

    Abs. Such variables as raw materials, mixing procedures, processing, and costs are discussed in this paper as they relate to high-strength premix.

74. Yovino, J. O., Boeker, B. E., and Torres, A. F., "A Study of the Variables Affecting the Corrosion Resistance of Premix," Proceedings SPI RPD 17th Ann. Conf., Chicago, IL, Feb. 6–8, 1962, Sec. 5-E (14 pp.).

    Abs. This paper discusses the effects of fillers, reinforcements and resin systems on the corrosion resistance of polyester premix.

75. Cutler, N. A., Ferriday, J. E., and Parker, F. J., "Recent Technical and Application Developments in DMC," Ibid., Sec. 5-F (16 pp.).

    Abs. The development of DMC in the U.K. and the principal differences between the U.K. and U.S. markets are discussed. A new crack-free improved DMC is mentioned.

76. Calderwood, R. H., and Hovanec, A. J., "Premix Mixing Procedures—A Study," Ibid., Sec. 5-G (4 pp.).

    Abs. Increasing the blade clearance in the mixer to ⅜ inch, mixing at elevated temperatures, and using HSI type glass all improve impact strength of a premix compound.

77. Scholtis, K., "Neuere uber Eigenschaften verarbeitung und Anwendung von Polyester-Prepregs," (Properties, Processing, and Applications of Polyester Prepregs) *Kunststoffe*, Dec. 1962, pp. 758–762.

    Abs. This paper presents a good discussion of the state-of-the-art of SMC development in West Germany. Sketches of SMC impregnators are included.

78. Calderwood, R. H., and Hovanec, A. J., "Premixing for High Impact Strength Reinforced Plastic," *Plas. Tech.*, Apr. 1962, p. 38.

Abs. Based on Item 76 above.

79. Burton, G. W., "An Evaluation of Flame Retardants for Polyester Laminates and Molding Compounds," *Plas. Des. & Proc.*, Aug. 1962, pp. 28–32.

    Abs. Flame-retardant test methods are discussed. Standard and halogenated polyester resins are evaluated as well as additives.

80. Jackson, R. J., "Premixes—A Solution to Many Design Problems," *Rein. Plas.*, Sept.–Oct. 1962, pp. 20–21.

    Abs. A description of premix compounds is given, and they are compared with other molding materials for electrical and physical properties. Some possible problem areas are mentioned.

81. Sutcliffe, M. R., "Degree of Cure in Polyester DMC," BPF 3rd RP Conf., London, Nov. 18–30, 1962, Paper DMC-3 (13 pp.).

    Abs. In this study DMC discs were molded at 110°C. and then subjected to various tests. Visual inspection of the molded discs was most successful. Hardness, power factor, and water absorption are considered confirmatory tests. Other tests showed no ability to determine degree of cure.

82. Milburn, N. A., "Mould Design For DMC," Ibid., Paper DMC-5, (19 pp.).

    Abs. This paper contains a general discussion of DMC mold design with several examples where problems occurred and what was done to solve the problems.

83. White, R. B., "DMC Applications in the USA," Ibid., Paper DMC-6 (21 pp., 17 photos).

    Abs. This paper deals with applications of premix compounds and is based in part on Item 71 above.

1963

84. Epel, J. N., "Mold and Part Design for Maximum Properties from Premix Molding Compounds," Proceedings SPI RPD 18th Ann. Conf., Chicago, IL, Feb. 5–7, 1963, Sec. 6-A.

Abs. A total of 11 design suggestions is offered to improve the rigidity and strength of premix parts.

85. Protrubacz, J. G., "Variables Affecting Release of Premix Compounds," Ibid., Sec. 6-D (6 pp.).

    Abs. A test method consisting of a small molded part is discussed. Data are presented on several release agents.

86. Rheinfrank, G., "Automatic Molding of Glass Reinforced Alkyd Molding Compounds," SPE RETEC, Cleveland, OH, Oct. 1–2, 1963, Preprints, pp. 1-n4.

    Abs. The operation of an automatic rotating cold plunger machine for BMC is described.

87. Mulder, A. C., "Reinforced Plastics Save Money and Improve Reliability of Welding Machines," *Plas. Des. & Proc.*, Jul. 1963, pp. 18–19.

    Abs. Parts molded of Glastic molding materials for welding machines are discussed.

88. Friedman, L. W., U.S. Patent 3,110,690, "Bivalent Metal Hydroxide Treatment of Drying Oil Modified Alkyd Resins," Nov. 11, 1963.

    Abs. Alkyd resin stabilized, but not solidified, by heating with oxides or hydroxides of Mg, Ca, Hg, or Pb.

89. Societe anonyme des sines Chausson, British Patent 941,481, Nov. 13, 1963.

    Abs. Unsaturated polyester resin pregelled with stannous chloride before being used to impregnate a glass mat.

1964

90. Anon., "Spotlight on Reinforced Plastics," *Plas. Des. & Proc.*, Jan. 1964, p. 22.

    Abs. The premix stock for the M-14 rifle is shown and discussed.

91. Savage, R. J., Irwin, T. J., and Austin, T. F., "Finishing of FRP Parts for Automotive Use," Proceedings 19th SPI RPD Meeting, Chicago, IL, Feb. 4–6, 1964, Sec. 2-A.

Abs. Much has been written and said about the problem of using very high-quality automotive finishes in sports car bodies such as Studebaker's Avanti and Chevrolet's Corvette. It is not realized by many people that there are numerous high-quality industrial finishes that may be applied very efficiently to reinforced plastic surfaces to provide decorative and highly protective surfaces. This paper reviews some of the parts that have been made at General Tire's Marion, Indiana, plant using alkyd baking enamels, acrylic lacquers, acrylic reflow lacquers, organosol coatings, and others.

92. Bassler, R. B., Caramante, D. E., and Diebold, W. R., "FRP Automotive Material, Process, and Finishing Improvements," Proceedings 19th SPI RDP Meeting, Chicago, IL, Feb. 4–6, 1964, Sec. 2-B.

    Abs. One of the basic requirements for full acceptance of any product, or for the product to reach its full growth potential, is development and application of its technology. With proper definition of problems, efforts can be concentrated on solutions and material improvements. For RP to be accepted in automotive use, physical properties and visual appearance after paint are important factors. A motivating factor in this activity has been to improve surface finish of FRP automotive grade laminates and thereby reduce finishing costs (authors).

93. Foster, L. P., "Methods of Reducing Premature Polymerization of Catalyzed Polyester Resins," Technical Paper presented to SPI 5th Premix Seminar, St. Louis, MO, April 16, 1964.

    Abs. This report discusses the use of several inhibitors to retard polymerization of polyester resins using the SPI gel characteristics test as the criteria. Information is provided for: HQ, MTBHQ, HQMME, DTBHQ, DTBPC, and MDTBP.

94. Epel, J. N., "Spiral Flow Mold for Pemix," Technical Paper presented to Premix Committee, Columbus, OH, Nov. 20, 1964.

    Abs. The Duralastic spiral flow test equipment is

described and the effects of mold temperature, catalyst type and concentration, and of several types of fillers are reported.

95. Potrubacz, J. G., "A New Exotherm Test Apparatus," Ibid.

   Abs. The Harrison Radiator block gel test procedure is described, and typical data are presented.

96. Yovino, J., Boeker, B. E., and Torres, A. F., "A Study of the Variables Affecting the Corrosion Resistance of Premix," Ibid.

   Abs. This report discusses such variables as resin type, monomer type, resin to monomer ratio, filler type, particle size, and coating, and reinforcement type, and surface treatment on the corrosion resistance of premix compounds.

97. Schweitzer, W. P., "Premix Extrusion and Preforming," Ibid.

   Abs. The advantages and disadvantages of clay type vs. screw type clay extruders for premixes are discussed.

98. Butler, J., "The Moulding Of DMC," BPF 4th RP Mtg., London, Nov. 25–27, 1964, Paper 12 (8 pp.).

   Abs. Molding problems with electrical insulation items are discussed. Mold design is well illustrated.

99. White, R. B., *Premix Molding*, Rheinhold Pub. Corp., New York, 1964 (201 pp.).

   Abs. The only exclusively premix book published. Regarding painting, internal and external mold release agents must be removed from a part before painting. Silicones must not be used on parts to be painted. Surface porosity in FRP parts cause painting problems.

100. Bassler, R. B., and Caramante, D. E., "New Fiberglass Reinforced Plastic Automotive Finishing Technique," OCF Report TR-228 (1964).

   Abs. This report describes the technique of measuring surface smoothness using the Bendix Microcorder, the proper sanding technique to overcome waviness, "bondburn," and the reduction of fiber pattern in laminates through modification of materials and molding technique.

1965

101. Eastwood, N., "Flomat Preimpregnated Chopped Strand Mat," Proceedings SPI RPD 20th Ann. Conf., Feb. 2–4, 1965, Sec. 6-A.

    Abs. Initial paper in the U.S. describing SMC products. Mold design is shown. Molding conditions and applications are discussed.

102. Savage, R. J., and Weiner, A. L., "Continuous Finishing System for FRP Automotive Parts," Proceedings 20th SPI RPD Meeting, Chicago, IL, Feb. 2–4, 1965, Sec. 10-A.

    Abs. The finishing of reinforced plastics parts divides into two separate but related areas of technology. One is the molding of surfaces on RP parts suitable for painting and finishing; the other is the actual painting and finishing operation. Our work at Marion has dealt extensively with both those areas, and much of the expansion that has taken place during the past year has been devoted to both facilities for molding and for finishing. A continuous finishing system for RP has resulted (authors).

103. Spiwak, L., "Thermoset Flow Tests," *Plas. Tech.*, Apr. 1965 (11-5), pp. 41–42.

    Abs. The ASTM flow cup, spiral flow, IBM, and Mesa special spiral flow test methods are described.

104. Schnitzler, J., "Glass Mats Preimpregnated with Polyester Resins (Prepregs)," *Plas. Verabiter* (16-4), Apr. 1965, pp. 171–178.

    Abs. An important article on SMC preparation showing equipment and formulations (in German).

105. Eshleman, L. H., "Premix Surface Finish Evaluation," Tech. paper presented at Premix Committee meeting, New York, May 13, 1965.

    Abs. This paper shows the effect of press variables on the surface finish of premix parts.

106. Ferrigno, T. H., "Effects of Fillers on the Surface Appearance of Molded Premix Parts," Ibid.

107. Eastwood, N., "Preimpregnated Chopped Strand Mat," *British Plas.*, Sept. 1965, pp. 559–563.

    Abs. Based on Item 101 above. *Abs.* The effect of particle size and type of filler on the surface appearance of premix parts is discussed.

108. Anon., "Impregnated Glass Fiber Mat Speeds Production," *Chem. & Engr. News*, Oct. 4, 1965, pp. 44–45.

    Abs. Initial Vibrin-Mat™ announcement.

109. Tiffin, A. J., "Structoform™—A New Molding Material For Improved FRP Processing," SPE RETEC, Cleveland, OH, Oct. 4, 1965, Preprints, pp. 68–69.

110. Walton, J. P., "Prethickened Polyester Resins in FRP Molding Compounds," Ibid.

    Abs. Initial Vibrin-Mat™ paper.

111. Kaull, G. H., "Pastel Premix or Out ¼ Million Pound Monthly Headache," Ibid., pp. 1–6.

    Abs. The problems involved in producing pastel-colored premix parts for the appliance industry are discussed.

112. Fesco, V., "Precipitated Calcium Carbonates as Fillers for Premix," Minutes SPI Premix Committee Meeting, Ashtabula, OH, Oct. 14, 1965.

    Abs. This paper discusses the various methods of manufacturing precipitated calcium carbonates, their properties, and their influence on cost and viscosity of premix compounds. Eight formulations are provided.

1966

113. Eshleman, L. H., "Premix Surface Finish Evaluation," Proceedings SPI RPD 21st Ann. Conf., Chicago, IL, Feb. 8–10, 1966, Sec. 6-B.

    Abs. Premix boses were molded at 100, 200, 300, 400, and 500 tons pressure and at mold temperatures of 250°F, 265°F, 280°F, and 300°F. Physical properties, surface smoothness, and shrinkage were evaluated. Parts molded at 300°F and 400 tons pressure had the best finish.

114. Tiffin, A. J., and Shank, R. S., "Molding Characteristics of Structoform-Reinforced Plastic," Ibid., Sec. 7-A (4 pp.).
    Abs. A description of the product, molding conditions, and typical physical properties is given.
115. Gruenwald, R., "Mold-Mat Preimpregnated Glass Mat—Its Comparison to the Preform and Premix Systems," Ibid., Sec. 10-D (8 pp.).
    Abs. A follow-up on Item 104 with considerable additional information.
116. Walton, J. P., "Mechanization of FRP Molding with Vibrinmat™," Ibid., Sec. 11-B (6 pp.).
117. Thomas, J. L., "DAPMIX™ for Improved Moldability of Premix," Minutes SPI Premix Committee Meeting, St. Louis, MO, Apr. 14, 1966.
    Abs. A new monomer for use in premix compounding is described together with its effect on reactivity and shelf life of the compound and on the physical properties of the mix. A number of formulations are provided.
118. Cramm, R. H., "Chlorostyrene in Premix and Mat Molding," Ibid.
    Abs. This paper discusses the use of the monomer O-chlorostyrene in polyester systems. It is compared with styrene and vinyltoluene in properties and the effect of various monomer-polyester ratios on physical properties is presented.
119. Walton, J. P., "Vibrin-Mix™, an Improved Premix Compound,"
    Abs. Prethickened polyester premix compounds are discussed.
120. Ensminger, R. I., "Pigment Dispersion in Synthetic Vehicles," Titanium Pigment Corporation, April 22, 1966.
    Abs. An excellent discussion of deflocculation, flocculation, and agglomeration stages of pigment dispersion. Also included is a discussion of dispersion equipment including low-vehicle-solids ball mill grinding, sand grinding, and high-speed-impeller mills.
121. Anon., "General Tire's Peppercorn Predicts Bright Future for Prepregs," *Autoproducts*, May 1966.

Abs. A question and answer interview by the editor of *Autoproducts* concerning Genite™ molding materials is presented.

122. Hersch, P., "What's New in Premix and Mold-Mat," *Plas. Tech.*, Aug. 1965 (12-8), pp. 31–34.

    Abs. The new raw materials that are available for premix are discussed as well as compounding notes and molding notes.

123. Ferstenberg, C., and Meyer, R. W., "Genite™ Precombined Materials," Preprints SPE RETEC, Cleveland, OH, Sept. 30, 1966, pp. 116–123.

    Abs. This paper describes Genite™ BMC and SMC materials.

124. Tiffin, A. J., "Molding Techniques for Structoform™ Reinforced Plastics," Ibid., pp. 98–104.

125. Walton, J. P., "The Physical Properties of Vibrin-Mat™ With Varying Molding Conditions," Ibid., pp. 105–115.

126. Morgan, D. E., and Sturman, D., "Assessment of the Heat Resistance of Polyester DMC," BPF 5th Conf., London, Nov. 23–25, 1966, Paper No. 6.

    Abs. This paper examines the rate of weight loss and the physical property retention of several DMCs subjected to elevated temperatures. The tests show that over a considerable life span of the materials there is essentially a linear relationship between weight loss and heating time at a given temperature.

127. Deis, W. H., "Marinco H Magnesium Hydroxide in Polyester Resins," Nov. 1966.

    Abs. Maturation studies involving five polyester resins are presented.

128. Anon., "Automated Prepreg Production," *Reinforced Plastics*, Nov.–Dec. 1966, p. 22.

129. Thomas, J., "DAPMIX for Improved Polyester Premix and Glass Mat Laminates," Preprints SPE RETEC, Chicago, IL, Dec. 12–13, 1966, pp. 1–10.

    Abs. DAPMIX is a DAP ester that is reactive with unsaturated polyester resins having thixotropic behavior that provides good surface finish and improves physical properties.

130. Hoffman, D. F., "Injection Molding of Thermosets," Ibid.
131. Anon., "Breakthrough in FRP," *Goodyear Aerospace Profile*, Vol. IV, No. 3.

Abs. Photographs of the original Goodyear SMC machine.

1967

132. Moore, C. E., and McMahan, J. D., "Paint Film Evaluation: Rohm & Haas Molding Formulation," Jan. 6, 1967, OCF Report No. TC/RPL/67/2.

    Abs. The Paraplex℗ AP-340-344 two-part system was found to save considerable time and materials since the dry sanding operation could be eliminated and less materials were required, as it was no longer necessary to obscure the fiber pattern.

133. Alspac, H. S., "Spraypreg—A New Engineering Material," SAE paper 670106, Jan. 9, 1967.

    Abs. Physical properties and current applications are described.

134. Walton, J. P., "Quality and Economy in Glass Reinforced Plastics with Vibrin-Mat Molding Compounds," Ibid.
135. Tiffin, A. L., Cunningham, R. G., and Waters, J. A., "Stuctoform Reinforced Plastics," Ibid.
136. Walton, J. P., "Molding Of Vibrin-Mat™—A Case Study," Proceedings 13th SPI RPD Ann. Conf., Washington, D.C., Feb. 1–3, 1967, Sec. 8-B, (6 pp.).

    Abs. Three grades of Vibrin-Mat™ are described, and eight typical molded parts are discussed.

137. Tiffin, A. J., Cunningham, R. G., and Waters, J. A., "Fabrication of Commercial Products with Structoform™ SMC," Ibid., Sec. 8-C.

    Abs. Structoform™ SMC is described with molding conditions and typical physical properties.

138. Nenadal, C. G., "FRP in Major Appliance-Premix-A Versatile Engineering Tool in Design Applications," Proceedings SPI RPD Ann. Conf., Washington, D.C.,

Jan. 31–Feb. 3, 1967, Sec. 15-C (6 pp., 11 photos).

Abs. Typical premix molded parts are illustrated.

139. Anon., "Fast Cure: Key to Three New Premix Resins," *Plas. Tech.*, Jan. 1967, p. 11.

Abs. Freeman Acpol™ resins are discussed.

140. Anon., "Stypol™ Prepreg Mat Resins," *Bulletin 17* RP-A&F, Feb. 16, 1967, Freeman Chemical Co., Pt. Washington, WI.

Abs. Product descriptions and recommended formulations for SMC are provided.

141. Tiffin, A. J., and Shank, R. S., U.S. Patent 3,305,514, "Vinyl Halide Resin Epoxy or Alkyd Resin, Monoalkyenyl Reinforced Thermoplastic Composition," Feb. 21, 1967.

Abs. Structoform™ patent.

142. Walton, J. P., "The Physical Properties of Vibrin-Mat™ with Varying Molding Conditions," *SPEJ*, Feb. 1967, p. 29.

Abs. Based on Item 136 above.

143. Eig, M., "Recent RP Progress: A Perspective," *Plas. Tech.*, Aug. 1967, p. 43.

Abs. New developments in premix molding are discussed, such as faster cure rates, better surface finish, etc.

144. Scholtis, K., Bergs, H., and Otromke, M., "The Flow Properties and Curing Characteristics of Glass Fibre Reinforced Polyester Moulding Compounds," *Kunststoffe*, Aug. 1967, pp. 635–637.

Abs. Glass fibre molding compounds can be tested by plotting the increase in kneading rersistance vs. time in a Brabender Plastograph. Curves are included to illustrate the effect on the curing characteristics of various factors, such as the nature and quantity of peroxide, the addition of inhibitors, storage time and curing temperature. Hundreds of measurements reveal that reproducible results can be obtained provided certain oprerating conditions are met. As the test takes only a few minutes, it can be used as a production control technique for polyester molding compounds.

145. Potrubacz, J., "The Harrison Radiator Polyester Premix Operation," Proceedings SPE RETEC, Cleveland, OH, Oct. 9–10, 1967.

    Abs. No paper in proceedings. Paper passed out at the meeting. This paper describes the premix program at Harrison Radiator to make heater housings for GM cars.

146. Walton, J. P., "Vibrin-Mat™ Sheet Molding Compounds—Key to FRP Automation," Tech. Paper, SPE RETEC, New Haven, CT, Nov. 1967, p. 94.

147. Walton, J. P., "Production Molding with Vibrin-Mat™," Tech. Talk, SPI Premix Comm. Mtg., Granville, OH, Nov. 9, 1967.

148. Eshleman, L. H. "Automatic Injection Molding of Acpol™ Fast Cure Premix Compounds," Ibid., Exhibit A.

    Abs. The use of a Seidl injection molding machine to mold premix compounds is described.

149. Campana, T., "The Effects of Soft Water on Premix Molding," Ibid., Exhibit C.

    Abs. The corrosive effects of ion exchange water softeners on premix compounds are discussed.

1968

150. Anon., "Injection Molder Speeds RP Molding," *Plas. Tech.*, Jan. 1968.

    Abs. The Seidl injection molding machine is described.

151. Moore, B. G., "Painting of Fiber Glass Reinforced Plastic," Proceedings 23rd SPI RPD Mtg., Washington, D.C., Feb. 6–9, 1968, Sec. 1-D.

    Abs. The painting development facilities at MFG are described. Boats are wiped with xylene and then painted. Automotive parts are sanded over 100% of their surfaces to remove fiber pattern and long-term waviness. The part then is putty rubbed to fill porosity. Then a 3–5 mil wet film thickness polyurethane primer is applied followed by a pink guide coat wet-on-wet, baked 30 min. @ 250°F (121°C).

152. Vaill, E. W., "Injection Molding of Thermosets," Proceedings SPI RP/CI 23rd Ann. Conf., Washington, D.C., Feb. 6–9, 1968, Sec. 11-A.

   Abs. The injection molding of thermosets requires equipment modifications, procedural changes, and mold design modifications as compared with thermoplastic molding. Process conditions are discussed, as are runner and gate design notes.

153. Grigor, Jr., J. M., "Reciprocating Screw Injection Molding Machine," Ibid., Sec. 11-B (4 pp., 8 photos).

   Abs. Minimum glass degradation is claimed for the screw, but degradation does occur because of the nozzle, gate, etc.

154. Paci, R., "Molding Reinforced Plastics with Screw Injection Equipment," Ibid., Sec. 11-C (2 pp.)

155. Kroekel, C. H., and Bartkus, E. J., "Low Shrink Polyester Resins—Performance and Applications," Ibid., Sec. 18-E.

   Abs. The Paraplex™ P-19A system is discussed.

156. Anon., "New Standard Spiral Flow Test," *Mod. Plas.*, Feb. 1968, pp. 104–108.

   Abs. The Epoxy Materials Institute developed a standard spiral test method. This paper gives the results of a study of molding conditions on the spiral flow length of an epoxy compound.

157. Anon., "Derakane™ Vinyl Ester Resins," Tech. Data Report No. 4, by Dow Chem. Co., Feb. 1968.

   Abs. Experimental vinyl ester resins QX-3923.40, QX-3923.50 and QX-3990.50 for chemically thickened molding compounds are described.

158. Kinnin, J. I., "The Effects of Cure Systems on Polyester Properties," Tech. paper delivered at SPI 13th Premix Committee Meeting, Galesburgh, IL, Apr. 4, 1968.

   Abs. The polyester ingredients, acids, glycols, and monomers are the main source of corrosion resistance. The catalyst system must be balanced to achieve the desired control over the cure mechanism.

159. Schnell, H., Raichle, K., Prater, K., and Brochne, F., U.S. Patent 3,390,205, Jun. 25, 1968, "Mono Carboxylic

Acids as Thickening Aids for Magnesium Oxide Containing Unsaturated Polyesters,"

160. Eshleman, L. H., "Acpol™ Premix Compounds—Seidl Injection Molding," SPE RETEC, Cleveland, OH, Sept. 30–Oct. 1, 1968, pp. 1–2.

Abs. Acpol™ premix compounds and the Seidl plunger injection molding machine are described.

161. Wright, F. J., "Low Shrink Sheet Molding Compound Vibrin-Mat™ L Series," Tech. Paper, SPE RETEC, Ibid., p. 63.

162. Blanchard, B., "How to Buy Premix Molding Equipment," *Plas. Tech.*, Oct. 1968, pp. 185–187.

Abs. This article is concerned with small FRP presses. Some mention is made of mixing equipment.

163. Anon. (OCF), "SMC Machine Manual and Operating Procedure," Jan. 27, 1968.

Abs. A description of the Granville SMC impregnator, operating instructions, chart of output vs. line speed and a troubleshooting guide is given.

1969

164. Walton, J. P., "Vibrin-Mat™ Sheet Molding Compound—A New Generation of Fiberglass Reinforced Polyesters," Tech. Paper, SAE Congress, Jan. 1969, paper 690010.

Abs. The preparation, manufacture, cost considerations and typical applications of SMC are discussed.

165. Hughes, G., "Low Cost Window Frames Using SMC," Proceedings SPI RP/CI 24th Ann. Mtg., Washington, D.C., Feb. 4–7, 1969, Sec. 3-A.

Abs. The history of the development of an FRP window frame project is discussed.

166. Frankenhoff, E. B., Maxel, J. M., and Miller, E. K., "Bulk Molding Compounds," Ibid., Sec. 3-B.

Abs. This paper discusses the latest developments in BMC with reference to chemically thickened systems and LP additives. Mixing techniques and charge preparation are included.

167. Nenadal, C. G., "Evaluations of Design with Premix in Appliance and Equipment," Ibid., Sec. 20-B.

    Abs. The design changes that occurred in humidifier and window air conditioners are discussed.

168. White, R. A., "Sheet Material Compound Aspect," Preprints SPI RP/CI 24th Ann. Meeting, Washington, D.C., Feb. 4–7, 1969, Sec. 8-C.

169. Zimmerman, A., and Wright, V., "What's Happening with SMC's?" Mos. Plas., pp. 36–40.

    Abs. General survey-type article.

170. Kwan, F. K., Ross, R. C., and Svoboda, G. R., U.S. Patent 3,432,458, "Thickened Polymerizable Unsaturated Polyester Resin Compositions and Fibrous Substances Impregnated Therewith."

    Abs. Unsaturated polyester thickened with alkaline earth oxide plus a polyhydric alcohol.

171. Dies, W., "Chemical Thickening Agents for Polyester Resins," presented at SPI Premix Committee Meeting, Grand Rapids, MI, Apr. 2, 1969.

    Abs. The variables that affect chemical thickening are discussed. Magnesium oxide and hydroxide and calcium oxide and hydroxide and their combinations are included.

172. Frankenhoff, E. B., Maxel, J. M., and Miller, E. R., "Developments in BMC Molding Compounds," Rein. Plas., May 1969, pp. 236–242.

    Abs. Based on Item 166 above.

173. Phillips, D. R., and Taylor, R. J., British Patent 1,155,119, "Moulded Articles," June 18, 1969.

    Abs. This patent teaches the use of chopped strand glass mat to reinforce DMC moldings.

174. Fekete, F., and Baum, M., U.S. Patent 3,465,061, "Unsaturated Resin Composition Having a Thickening Agent Therein," Sept. 2, 1969.

    Abs. Group II metal oxide and cyclic hydrocarbon anhydride are used to thicken unsaturated polyesters.

175. Jerigan, J. W., U.S. Patent 3,466,259, Sept. 9, 1969.

Abs. Claims thickening of vinyl ester resins with Group II metal oxide or hydroxide and water.

176. Kent, J. R., "Injection Molding Techniques for Polyester Premix Processing," Preprints SPE RETEC, Cleveland, OH, Sept. 25, 1969, pp. 1–5.

Abs. A reciprocating screw injection molding machine with a nonreturn valve is described.

177. Cordts, H., "One Component Low Shrink System," Ibid., pp. 67–69.

Abs. A product announcement on Stypol™ 40-7000 for premix and wet molding systems.

178. Walton, J. P., "The ABC's of Making SMC," Ibid., pp. 55–73.

Abs. A good review of raw materials, maturation, and of processing SMC.

179. Foley, N. J., "An Advanced Design System for Mixing FRP," Ibid., pp. 79–84.

Abs. This paper discusses a new BMC muller-type mixer.

180. Fekete, F., "Latest Developments in SMC," Ibid., pp. 84 (44 pp.).

Abs. Abstract only in proceedings. Paper passed out at the meeting. Thickening rate studies of various Group II compounds are presented.

181. Canning, J. L., and Sias, C. B., "Rheological Properties of Unsaturated Polyester Molding Compounds," Ibid., p. 88.

Abs. Abstract only in proceedings. Paper passed out at the meeting. The various flow tests used on premix are discussed. A new flow tester is detailed. No data are included.

182. Handlovits, C. E., and Anderson, T. E., "Low Shrink Premix," Ibid., p. 97.

Abs. Abstract only in proceedings. Paper passed out at the meeting. This paper describes the use of t-butyl styrene in premix compounding.

183. Beck, Koller & Co., British patent 1,165,622, Oct. 1, 1969.

Abs. An alpha beta unsaturated dicarboxylic acid with an oxide, hydroxide, or salt of an alkaline earth metal is claimed to shorten the maturation period of SMC.

184. Pfaff, F., and Bartkus, E., "Sink Mark Formation in Parts Molded From P-19C Molding Compounds," Oct. 22, 1969.

    Abs. A study using the OCF rib mold.

185. Anon., "Low Shrink BMC," Tech. Data Sheet published by Koppers Co., Oct. 1969 (4 pp., 3 formulations).

    Abs. This data sheet describes recommended formulations for LS BMC, mixing instructions, and typical properties for these formulations.

186. Wright, F. M., "Low Shrink (Low Profile) SMC," Thermoset Molding Processes for Tomorrow, The University of Wisconsin Extension Seminar, Oct. 1969.

187. Davies, E. A., "Polyester Molding Compounds for Automotive Parts," Preprints RP Symposium, BPF, Plastics In Motor Cars, Solihull, U.K., Nov. 12, 1969.

    Abs. A review of the application of DMC for automotive exterior parts for the 1969 model car year in the U.S. The use of chopped strand mat to reinforce DMC is outlined, as is the effect of temperature on the physical properties of DMC molded parts.

188. Stoops, R., and Maxel, J., "Smoother, Better-Looking Reinforced Plastic Molding," *Materials Engr.*, Dec. 1969.

    Abs. Rohm & Haas formulations are discussed and Mercury and Pontiac parts are shown.

189. Tiffan, A. J., "Sheet Molding Compound-Strength vs. Versatility," *Rein. Plas. & Composites World*, 1969 annual issue, p. 48.

1970

190. Stone, D. H., "Injection Molding of Polyester Premix," Proceedings SPI RP/CI 25th Ann. Mtg., Washington, D.C., Feb. 3–6, 1970, Sec. 1-B.

Abs. It is claimed that P-19A premix when injection molded has increased automation, lower labor cost, faster cycle times, and improved part quality.

191. Fekete, F., "Thickeners and Low Shrink Additives for Premix and SMC Systems," Ibid., Sec. 6-D.

    Abs. This paper involves an extensive study of chemical thickeners and thickener accelerators on Koppers polyester resins.

192. Nussbaum, H. W., and Czarnomski, J. J., "Smooth Surface Premix and SMC," Ibid., Sec. 6-E.

    Abs. The majority of this paper is concerned with SMC. Several thickeners are discussed with MgO being preferred. Fillers, glass, and molding details are also mentioned.

193. Carter, N. A., "Sheet Molding Compounds, Part 1: Formulations, Processing, Part Design," *Plas. Des. & Proc.*, Feb. 1970, pp. 18–21.

194. Anon., "Plastics in Transportation: Slower Car Sales Won't Hurt Growth," *Plas. Tech.*, Feb. 1970.

    Abs. Statistics on the use of FRP in trucks, buses, and cars are briefly described.

195. Anon., "Sheet Molding Compounds, Part 2: Presses, Molds," *Plas. Des. & Proc.*, Mar. 1970.

196. Tagaki, S., Yokoo, M., and Hatanaka, Y., U.S. Patent 3,503,920, Mar. 31, 1970.

    Abs. Magnesium carbonate is claimed to thicken polyester resins.

197. Anon., "SMC Going In-Plant," *Mod. Plas.*, May 1970, pp. 68–70.

    Abs. Photos of the OCF Granville SMC impregnator.

198. Davies, E. A., "Non-Shrink or LP Molding compounds for the Automotive Industry," BPF RP Conf., London, May 1, 1970 (41 pp.).

    Abs. The significant properties of LP-DMC are listed as: retention of properties over a wide temperature range; nonshrink properties that provide excellent surface finish; close tolerance moldability, low coefficient of thermal expansion; and good resistance to hostile environments.

199. Anon., Cellobond™ K-515 Tech. Data sheet, published by BP Chemicals Ltd., London (8 pp.).

Abs. This data sheet shows some automotive applications and illustrates the effect of temperature on flexural strength, flexural modulus, tensile strength, and impact strength of DMC K-515.

200. Kallaur, M., German Patent 1,943,181, Jul. 30, 1970.

Abs. A mixture of polyester, vinyl monomer, catalyst, and inhibitor is thickened with CaO or Ca(OH)$_2$ and water.

201. Espenshade, D.T. "Injection Molding of Polyester Resins," Proceedings SPE RETEC, Cleveland, OH, Sept. 28–30, 1970, pp. 1–24.

Abs. The Paraplex™ P-19A premix system is discussed with emphasis on injection molding. Considerable information is presented on studies of gate and runner design and gate location.

202. Carter, N. A., "Applications and Design Trends," Ibid., pp. 120–122.

Abs. Essentially a slide talk of FRP accomplishments and suggestions for future products.

203. Blue, E. G., "SMC Equipment Design," Ibid., pp. 126–129.

Abs. A description of the E. B. Blue SMC machine is provided. There are no sketches or photographs.

204. Howell, D. M., "Current SMC Automotive Applications," Ibid., pp. 130–132.

Abs. A slide talk of SMC applications. No photos in proceedings.

205. Cleaver, S. M., "Developments in The Injection Molding of DMC," *British Plas.*, Sept. 1970, pp. 115–118.

Abs. The Stokes Injectoset screw transfer and horizontal inline premix presses are described and illustrated. Molding conditions and test results using different sprue locations are given.

206. Vansco-Szmercsanyi, I., "The Interaction Between Unsaturated Polyester Resins and Metal Oxides," Hungarian Res. Inst., Oct. 6–9, 1970.

Abs. It is proposed that two reactions occur in sequence.

First, a salt is formed, secondly, a coordinate bonding is formed when certain metal oxides are added to −COOH containing polyester resins. (See discussion on mechanism of thickening in chapter 1.)

207. Anon., "SMC Plastic Called Material of Future," *Automotive Industries*, Oct. 15, 1970.

Abs. A news release on Houdaille's plans to enter the FRP market as a custom molder.

208. Raffel, B., Sheatsley, R., U.S. Patent 3,535,151, Oct. 20, 1970.

Abs. Rapid thickening polyester system obtained by spraying a mixture of polyester and MgO, Mg(OH)$_2$ or Ca(OH)$_2$ onto glass strands followed by spraying controlled quantities of water.

209. Fekete, F., "Thickeners for RP Sheet and Bulk Molding Compounds," *Mod. Plas.*, Oct. 1970, pp. 154–158.

Abs. A slow initial viscosity build up of 12–24 hours followed by a rapid viscosity increase when CaO is used in systems containing Ca(OH)^2 or MgO.

210. Anon., "Fiberglas/Plastic Design Guide," Publication No. 14-AV-5164 by OCF, Toledo, OH, Oct. 1970.

Abs. General process description, special and general design guides for SMC, BMC, and wet process and injection molding are provided.

211. Fekete, F., and Baum, M., U.S. Patent 3,538,188, Nov. 3, 1970.

Abs. The initial thickening rate of polyesters containing MgO or Mg(OH)$_2$ is slowed by the addition of lithium chloride, bromide, or nitrate.

212. Miller, E. R., "What to Look for in SMC Production Machines," *Plas. Tech.*, Dec. 1970, pp. 23–27.

Abs. SMC impregnator components are described. The Baker-Perkins continuous mixing system is discussed.

1971

213. DePalma, E. P., and Lehmkuhl, D. E., "Automotive Developments in RP," SAE Paper 710022, Jan. 11–15, 1971, Detroit.

Abs. A short history of FRP in automobiles is presented followed by an illustrated discussion of 1971 model year applications.

214. Norman, M. K., "Low Profile, HMC and SMC Applications and Design Guidelines," Ibid., paper 710023.

    Abs. The 1970–1971 model year automotive applications at Guide Lamp are discussed. Proper part design is stressed, and attachment of parts to automobiles can be a problem.

215. Epel, J. N., "Comparison of Low Profile Premix and SMC Applications," Ibid., paper 710024.

    Abs. BMC and SMC materials and manufacturing methods are described and physical properties compared. Some typical applications are described.

216. Dalhuisen, A. J., U.S. Patent 3,557,042, Jan. 19, 1971.

    Abs. This patent covers thickening polyester resins with anhydrous calcium propionate, calcium laurate, zinc propionate, and zinc salicylate.

217. Anon., "Planning High Production SMC?—Try This New Continuous Resin Mixing System," *Plas. Tech.*, Jan. 1971, p. 11.

    Abs. Baker-Perkins Rotofeed™ continuous mixer is discussed.

218. Potkanowicz, E. J., U.S. Patent 3,560,294, "Method and Apparatus for Combining a Viscous Resin and Glass Fiber Strand," Feb. 2, 1971.

    Abs. Patent covering a roll type SMC impregnator.

219. Kressin, D. M., and Kolezynski, J. R., "The Use of Organic Peroxides in Polyester Molding," Proceedings, SPI RP/CI 26th Ann. Mtg., Washington, D.C., Feb. 9–12, 1971, Sec. 4-D.

    Abs. A study of the effects of peroxide concentration and molding temperature on cure properties using 7 organic peroxides is reported.

220. McCluskey, J., and Bell, R. Z., "No Sink SMC-Status Report" Ibid., Sec. 6-C.

    Abs. A method of evaluating sink marks is presented, and procedures for obscuring sink marks are discussed. (Note: Paper was passed out at the meeting.)

221. Espenshade, D. T., and Lowry, J. R., "Low Shrink Polyesters for SMC," Ibid., Sec. 12-F, (10 pp.).

Abs. A discussion of the Paraplex™ P-19C SMC system is presented. Preferred formulations are shown and molding suggestions offered.

222. Sheatsley, R. W., and Ring, E., "SMC Controls for Quality Moldings," Ibid., Sec. 15-B.

Abs. A very general article recommending controls on raw materials, resin mixing, SMC preparation, and molding parameters.

223. Keown, J. A., "Fiberglas™/Plastic Design Guide," Ibid., Sec. 16-E.

Abs. This paper includes a description of SMC, specific part and mold design recommendations and a table of general design criteria.

224. Frick, D. K., "Innovations in Press Design for SMC," Ibid., Sec. 18-A.

Abs. A large hydraulic press especially designed for SMC is described.

225. Von Tesmar, C., "Finally, One Year Later—Commercial SMC," Ibid., Sec. 18-B.

Abs. The problems of developing an SMC for sale to industry are described.

226. Meyer, R. W., "Automation in Press Molded FRP," Ibid., Sec. 21-A.

Abs. The development of the SMC impregnator is discussed, and four sketches of typical designs are shown.

227. Walling, S. J., "Reinforcements," Ibid., Sec. 21-#.

Abs. A tabulation of SMC rovings is given together with roving characteristics.

228. Walton, J. P., and Wright, F. M., "Sheet Molding Compounds," Ibid., Sec. 21-F.

Abs. The history of SMC development is discussed through 1970 (52 references).

229. Keown, J. A., "Designing With SMC," *Mach. Des.*, Feb. 18, 1971, pp. 111–113.

230. Anon., "SMC Evaluation Project 0245-44", Tech. Report, Freeman Chem. Co., Pt. Washington, WI, Apr. 14, 1971.

  Abs. The results are reported on a 10-run SMC evaluation involving 3 Freeman resins and 4 thermoplastic additives.

231. Walker, A. C., "Some Mechanisms of Low-Profile Resin System Behavior," 29th SPE ANTEC, May 10–13, 1971, pp. 454–456.

232. Espenshade, D. T., and Lowry, J. R., "Low-Shrink Polyester Resins for SMC," Ibid., pp. 457–463.

233. Alvey, F. B., "Study of the Reaction of Polyester Resins With Magnesium Oxide," *J. Polymer Science*, Sept. 1971, pp. 2233–2245.

  Abs. The reaction of polyester resins and MgO were studied at three different temperatures. The effect of concentration and added water are reported.

234. Busch, W., and Schulz, H., "Schrumpfarme Polyesterformassen: Chemie, Technologie und Andwendungspeispicle" (LS Polyester Molding Compositions: Chemistry, Technology, and Examples of Usage), *Kunststoffe*, pp. 602–609.

  Abs. An extensive discussion of the chemistry of LP systems with numerous photomicrographs.

235. Davis, Sr., C. J., Wood, R. P. and Miller, E. R., U.S. patent 3,615,979 assigned to Owens-Corning Fiberglas Corp., Toledo, OH.

236. Williger, E. J., U.S. Patent 3,536,642, Oct. 27, 1971.

  Abs. The viscosity increase during storage or aging of polyesters can be reduced by the addition of hydroxylated compounds and salts of Al and Ga.

237. Anon., "Is This Anytime to Become a Custom SMC Molder?" *Mod. Plas.*, Nov. 1971, pp. 40–43.

  Abs. The story of Engineering Molding Systems Div., Houdaille Industries, Inc.

1972

238. Tudor, W. E., "Some Considerations Relative to Use of SMC for Automotive Exterior Body Panels," SAE paper 720061, Detroit, MI, Jan. 10–14, 1972.

Abs. The advantages and limitations of FRP body panels are discussed.

239. Stavinoha, R. F., and Macrae, J. D., "Derakane™ Vinyl Ester Resins Unique Chemistry for Unique SMC Opportunities," proceedings SPI 27th Ann. Conf., Washington, D.C., Feb. 8–11, 1972, Sec. 2-E.

    Abs. A brief description of vinyl ester chemistry is given.

240. Bradish, F. W., "Paint and Finishing Considerations for Low-Profile Systems," Ibid., Sec. 6-E.

    Abs. Abstract only in proceedings.

241. Grotke, M., and Garst, J., "Direct Digital Computer Control of Polyester Resin Batch Manufacturing Process," Ibid., Sec. 10-A.

    Abs. The development of a digital computer control system for the production of unsaturated polyester resins is described.

242. Ampthor, F. J., and Kroekel, C. H., "Development in LP-SMC for Flame Retardant and Electrical Applications," Ibid., Sec. 8-E.

    Abs. The use of Paraplex™ P-19C with alumina trihydrate and without flame retardants is discussed.

243. Leichtle, I. J., "Low Profile Systems Design," Ibid., Sec. 10-D.

    Abs. Abstract only in proceedings.

244. Wright, F. J., "New, 'No Shrink' Polyester Resin for SMC," Ibid., Sec. 12-A.

    Abs. The properties of GR-63007 polyester resin are discussed.

245. Rosato, D. V., "Racetrack Molding System for SMC," Ibid., Sec. 12-B.

    Abs. A single-compression press applies initial pressure after which clamps retain partial pressure on the molds as they move along a circular track, which holds up to 12 electrically heated molds.

246. Blount, W. W., Calendine, R. H., and Davies, J. H., "Cellulose Acetate Butyrate Polymers as LP Additives for SMC and BMC," Ibid., Sec. 12-C.

Abs. Three CAB esters that form one component LP systems with polyester are described.

247. Fekete, F., "A Review of the Status of Thickening Systems for SMC, LS-SMC, BMC, and LS-BMC Compounds," Ibid., Sec. 12-D.

    Abs. A comprehensive review of the chemical technology of various SMC and BMC systems is provided, paying particular attention to thickening.

248. Cutshall, J. E., and Pennington, D. W., "Vinyl Ester Resins for Automotive SMC," Ibid., Sec. 15-A.

    Abs. Vinyls ester resins are claimed to solve the SMC problems contributed by polyesters.

249. Hirano, H., and Yotsuzuka, M., "SMC in Japan," Ibid., Sec. 15-B.

    Abs. The history of FRP in Japan with particular emphasis on SMC materials is discussed. Considerable statistical comparison data are presented.

250. McNally, J. S., Dzik, C. J., and Williams, D. R., "Low Shrink Plus Flame Retardance," Ibid., Sec. 15-C.

    Abs. Abstract only in proceedings.

251. Ring, E., "An SMC Product from Concept to Market," Ibid., Sec. 15-D.

    Abs. A very general discussion on the handling of a customer order for a machine housing.

252. Rabenold, R., "Sink Factors in Low Profile Molding," Ibid., Sec. 15-E.

    Abs. Several factors that affect sink marks in molded parts are covered, including chemical and thermal shrinkage, the low-profile mechanism, and the effects of rheology and compression.

253. Maaghul, J., "SMC-Optimization of Reinforcing Fiber Glass," Ibid., Sec. 15-G.

    Abs. Abstract only in proceedings. Paper passed out at meeting.

254. Parker, F. J., and Nicklin, N. B., "The Decoration of BMC and SMC Mouldings by a Foiling Process," Ibid., Sec. 16-A.

    Abs. A simple 2-stage molding process is described. The second stage involves molding a glaze or surface coating over a previously molded part.

255. Anon., "Amoco IPA in LP Resin Improves SMC and BMC," Bulletin 1P-31b pub. by Amoco Chem. Co., Chicago, Feb 1972.

Abs. This bulletin contains the formulas and cooking schedule for resin SG-30 and a BMC formulation. Several low-profile additives are suggested.

256. Rizzi, M., "Total System Approach to Mixing and Injection Molding Glass Reinforced Polyester," Proceedings, SPI RP/CI meeting, Washington, D.C., Feb. 8–11, 1972, Section 16-D.

257. Kerr, R. C., "Tooling for SMC Compression Molding," *Mod. Plas.*, pp. 68–70.

Abs. The use of heel blocks, 0.002–0.003 inch shear edge clearances, and ⅜–½ inch telescope are recommended.

258. Kroekle, C. H., U.S. Patent 3,642,672, "Unsaturated Polyester Resinous Composition Containing Cellulose Esters and Molded Articles Therefrom," May 15, 1972.

259. Currier, G. J., "Development of the first SMC Body Component: the Chrysler Station Wagon Air Deflector," SAE paper 720493, May 22–26, 1972, Detroit.

Abs. The history of the Chrysler air deflector is reviewed.

260. Blount, W. W., "CAB as LP Additives for SMC and BMC," *Mod. Plas.*, May 1972, pp. 68–72.

261. Parker, F. J., and Nicklin, M. B., "Decorating BMC and SMC Moldings by a Foiling Process," *Mod. Plas.*, Aug. 1972, pp. 62–63.

Abs. Two-stage molding process based on Item 254 above.

262. Sieglaff, C. L., "Use Flow Properties to Improve Bulk and SMC Operations," *Plas. Des. & Proc.*, Nov. 1972, pp. 16–18.

Abs. The rheological properties of filled resin-thermoplastic additive mixtures are discussed.

263. Anon., "Fiberglas™/Plastics Design Guide," Pub. No. 5-AU-5164b, pub. by Owens-Corning Fiberglas Corp., Toledo, Nov. 1972.

Abs. An illustrated description of SMC and BMC is furnished together with specific part and mold design criteria. The latest information on minimizing sink marks is also included.

264. Kroekel, C. H., "Unsaturated Polyester Resinous Compositions," U.S. Patent 3,701,748, Oct. 31, 1972.

265. Anon., "Where Exactly Do We Stand on SMC's Today?" *Mod. Plas.*, pp. 42–44.

   Abs. Photographs of the Oldsmobile plastics operation are included.

266. Anon., "Large-Piece Molding Opens New Plastics Markets," *Plas. World*, Dec. 1972 (3 pp.).

   Abs. A general description of the molding facilities of General Tire & Rubber Company at Ionia, MI and Marion, IN.

1973

267. Jutte, R. B., "Sink-Mechanisms and Techniques of Minimizing Sink," SAE paper 730171, Jan. 8–12, 1973.

   Abs. Two major causes of sink marks over ribs and bosses are polymerization shrinkage and thermal shrinkage. Minimum lead-in radii are recommended as well as short glass fibers and increased glass loading. Best results occurred with continuous strand mat near the surface and ½ inch SMC in the rib area.

268. Drew, E. W., "Painting of DMC and SMC," *Rein. Plas.* (17-1), Jan. 1973, pp. 13–14.

   Abs. The advantages of painting DMC and SMC rather than molding in color are discussed.

269. Anon., "SMC Reduces FRP Costs and Raises Strengths," *At. Engr.*, Jan. 1973, pp. 38–41.

   Abs. SMC properties are compared with BMC. Better tensile strength is claimed for SMC.

270. Atkins, K. E., Harold, M. A., Comstock, L. R., and Smith, R. L., "Vinyl Polymers and Caprolactone Polymers as LP Additives," Proceedings SPI RP/CI Mtg., Washington, D.C., Feb. 6–9, 1973, Sec. 1-A.

   Abs. The use of polyvinyl acetate, vinyl acetate-vinyl

chloride copolymers, and caprolactone as LP additives is discussed.

271. Bradley, R. H., and Huminiski, F. M., "A New Unsaturated Polyester Resin for Color Applications," Ibid., Sec. 2-B.
Abs. The Paraplex™ P-19C system is discussed.

272. Seymour, M., "Design Research of an SMC Tub/Shower," Ibid., Sec. 7-E.
Abs. Only the abstract is included in the proceedings.

273. Green, D. H., and Guervin, P. R., "Polyurethanes—The New FRP Finish," Ibid., Sec. 8-B.
Abs. This article covers the types of polyurethane coatings, the diisocyanates used in their manufacture, surface preparation of FRP parts before painting, application procedures, and exterior and interior applications for these coatings.

274. Keown, J. A., "SMC Versatility in the Design Construction of GM's '73 Motor Home," Ibid., Sec. 12-D.
Abs. The history and design of the exterior panels for the GM 1973 Motor Home is reviewed.

275. Clavadetcher, D. J., English, F. G., and McGill, T. A., "New Market Applications in SMC and BMC Demand Precision Molding," Ibid., Sec. 18-A.
Abs. Some guide lines for tool design and molding conditions to obtain precision molded parts are given.

276. Luchini, A., "New Pigmented SMC for the Furniture Industry," Ibid., Sec. 18-B.

277. Ring, E., "SMC Capabilities—A Five-Year History at Goodyear Aerospace Corporation," Ibid., Sec. 18-C.
Abs. The history of SMC materials at GAC is discussed.

278. Horton, A. S., "Design of Computer Housings With SMC," Ibid., Sec. 18-D.

279. Selley, H. E., Shah, N. N., and Kay, D. J., "Formulating FR SMC," Ibid., Sec. 18-E.
Abs. Only the abstract is included in the proceedings.

280. Conley, D. O., "Status of Pigmented LS SMC and BMC," Ibid., Sec. 19-A.
Abs. The OCF E-930 resin system is discussed.

281. Austin, T. F., Tropp, F. E., and Pulman, L. J., "Continuous Metering and Process Control of SMC," Ibid., Sec. 19-D.

   Abs. The equipment used on the Koppers lab SMC unit is described, and 10 raw materials and in process test methods are presented.

282. Warner, K. N., "Mechanism of the Thickening of Polyester Resins by Alkaline Earth Oxides and Hydroxides," Ibid., Sec. 19-E.

   Abs. A 2-stage mechanism is proposed. A salt formation is followed by a diffusion reaction.

283. Fekete, F., "New and Novel SMC and BMC Compounds," Ibid., Sec. 19-F.

   Abs. A review of the new resins, new thickening techniques, and LP additives is presented together with formulations using the new materials.

284. Wood, A. S., "Detroit Isn't Your Only Low Shrink Market," *Mod. Plas.*, Feb. 1973, pp. 58–61.

   Abs. Applications of SMC in appliances, furniture, electrical, industrial components, and for corrosive service are discussed.

285. Pattison, V. A., Hindersin, R. R., and Schwartz, W. T., "Mechanism Of LP Behavior In Polyester Systems," SPE ANTEC, Montreal, May 1973, pp. 553–558.

   Abs. This paper is concerned with the resin system and the mechanism by which LS and LP behavior occurs.

286. Atkins, K. E., Harold, M. A., Comstock, L. R., and Smith, P. L., "Vinyl Polymers and Caprolactone Polymers as LP Additives for Unsaturated Polyester Molding," SPE 31st ANTEC, Montreal, May 7–10, 1973, pp. 527–531.

   Abs. The use of polyvinyl acetate and caprolactone polymers and copolymers of vinyl acetate and vinyl chloride as LP additives are discussed.

287. Pattison, V. A., Hindersinn, R. R., and Schwartz, W. T., "Mechanism of LP Behavior in Unsaturated Polyester Systems," Ibid., pp. 533–538.

   Abs. Photomicrographs made on a hot stage follow the curing process and are used to explain the possible mechanisms involved.

288. Deis, W. H., and Ness, W. W., "Magnesium Oxide and Magnesium Hydroxide for SMC," SPI RTPMC 5th Meeting, Huntsville, AL, June 6, 1973.

Abs. The properties of MgO and Mg(OH)$_2$ that affect their use as a chemical thickening agent in SMC are discussed.

289. Seymour, M., "Design of a New SMC Tub/Shower," Ibid.

Abs. The history of the development of OCF's tub/shower is related. Profusely illustrated.

290. Hall, A., "Prime Mover in Plastics-Oldsmobile," *Plas. World*, Sept. 17, 1973, pp. 41–54.

291. Tortolano, F. W., "With SMC Production Is a Whole New Ball Game," *Plas. World*, Sept. 17, 1973, pp. 62–64.

Abs. The molding of SMC bathtubs at the OCF Huntsville facility is discussed.

292. Meyer, R. W., "How to Mold SMC," N.Y. University 1st SMC Conference, Miami, FL, Oct. 11–12, 1973.

Abs. The variables involved in molding SMC are discussed. The use of the Audrey™ dielectrometer is reviewed as well as mold design criteria.

293. Morrison, R. S., "A Preliminary Evaluation of SMC and Mat and Preform Reinforced Plastics," Ibid.

Abs. The results of a molding study comparing SMC with mat and preform reinforcements in a concrete dome part are presented.

294. Walton, J., "SMC—When to Go Captive," Ibid.

Abs. The SMC market picture is presented as well as representative costs for several situations. General rules for whether to make or buy SMC are given.

295. Anon., "Evaluation of the Corvette FRP Body," *Corvette News*, Oct.–Nov. 1973, pp. 8–16, Chevrolet Motor Div. of GM.

1974

296. G. B. Lewis Co., German Patent 2,405,495, "Uncured GRP Dough Sheet Made by Continuous Process with

Double Folded Polyethylene Film Envelope," Feb. 5, 1974.

Abs. A single wide ply of PE film is used on the bottom of the sandwich. Forming dies fold the film edges to form an envelope.

297. Uffner, M. W., "Novel Promoters for Peroxyester Initiated Cures of SMC," Proceedings SPI RP/CI 29th Ann. Conf., Washington, D.C., Feb. 5–8, 1974, Sec. 1-B.

Abs. The use of promoters to reduce the cure time of SMC molded parts is discussed.

298. Woodhaua, R. T., and Xantos, M., "Unsaturated Polyester Resins Reinforced With HAR Mica," Ibid., Sec. 4-E.

Abs. Powder blending is preferred to liquid blending for high loadings of HAR compounds. Special resins are required. The use of silane coupling agents improves physical properties.

299. McKenzie, F. M., and Mauro, B. E., "Two Component Urethane Finishing Systems for RP Substrates," Ibid., Sec. 5-A.

Abs. This paper discusses the 2-component polyurethane finishing systems for SMC and plastics in general. Texture and high gloss systems for each type of substrate are listed together with a description of primer system, cure schedules, flash times, solvent limitations, and associated operations.

300. Selley, J. E., Kay, D. J., and Richards, W. B., "SMC for Corrosion Resistant Applications," Ibid., Sec. 9-A.

Abs. Abstract only in the proceedings.

301. Forsyth, G. E., and Ampthor, P. J., "Improved Maturation Control of LP SMC," Ibid., Sec. 9-B.

Abs. The use of Paraplex™ CM-201 as an additive to control the maturation of SMC pastes is discussed.

302. Maaghul, J., "SMC-Cost and Performance Optimization," Ibid., Sec. 9-C.

Abs. The use of statistical analysis to optimize some SMC variables is discussed.

303. Meyer, R. W., "Determining Cure Time of FRP Panels by Dielectrometry," Ibid., Sec. 9-E.

Abs. The use of continuous dielectric monitoring to determine the exact moment of cure of a molded panel is discussed.

304. Fekete, F., "Maximized Self Extinguishing LS-SMC," Proceedings, SPI RP/CI 29th Ann. Meeting, Washington, D.C., Feb. 5–8, 1974, Sec. 9-F.

Abs. Abstract only in proceedings.

305. Seymour, M. W., "SMC Plumbing Components and Systems Design," Ibid., Sec. 12-C.

Abs. The case history of the design of bathroom fixtures molded of SMC is detailed.

306. Bauer, S. H., "Injection Molding BMC," Ibid., Sec. 15-B.

Abs. Screw vs. plunger injection molding techniques are discussed. Problems with screw molding are detailed. The development of the stuffer hopper minimized labor input.

307. Milton, A., and Jestin, J., "Development of Continuous SMC Process," Ibid., Sec. 19-A.

Abs. A brief description of the Rhone-Progil SMC machine is given.

308. LeMay, R. A., Ibid., "SMC in Office Mailing Equipment—A Study in Part Consolidation," Ibid., Sec. 19-D.

309. Leichtle, I. J., "Applications of SMC Technology," Ibid., Sec. 22-D.

Abs. A picture plant tour of a Rockwell International molding plant is provided with emphasis on the production of Ford pickup truck covers.

310. Burden, J., "Single Component LP Polyester Resins," Ibid., *Olym. Age.*, (5-2/3), pp. 43–56.

Abs. New grades of BMC and SMC based on Synolac™ 6401 are described.

311. Wright, F. M., "Current Developments in SMC," SPE RETEC, Philadelphia, PA, Mar. 5–6, 1974, Preprints, Sec. 13.

Abs. SMC's are available in general purpose, low-shrink, electrical, flame-retardant, chemical, and weather-resistant grades. Molding technique is similiar to premix, but physical properties are comparable to mat or preform moldings. Growth of SMC will continue to be greater than for FRP in general.

312. Anon., "SMC Makes Computer Terminal for Less," *Plas. World*, Mar. 18, 1974, pp. 62–65.

    Abs. Computer bases of SMC and covers of BMC by Duralastic for Burroughs Corporation are shown.

313. Anon., "Now, SMC with Built-In Color," *Plas. World*, May 20, 1974, pp. 68–70.

    Abs. A description of the Hawley color molded-in SMC material and process is provided.

314. Miller, B. S., "97% Yield in Molding Thermosets?—Cold Manifold Molds Can Do It," *Plas. World*, Jun. 17, 1974, pp. 53–54.

    Abs. Cold manifold injection molded thermosets accomplish what hot runners do for thermoplastics: save material that otherwise would end up as sprue and runner scrap that cannot be recycled.

315. Walton, J., "When to Go In-House in SMC—If at All," *Plas. Tech.*, Jun. 1974, pp. 45–47.

    Abs. Three steps to SMC processing are discussed: first buying molded parts, then molding purchased SMC, and finally making and molding your own compound. A good cost breakdown is provided.

316. Dreger, D. R., "Which Processes for FRP Parts?" *Arch. Des.*, Sept. 19, 1974, pp. 128–133.

    Abs. A guide to the selection of FRP processes is provided.

317. Schicktanz, R., "Beitrag zur Herstellung von Harzmatten," (Contribution to the production of SMC), *Plasverbeiter*, Sept. 1974, pp. 555–562.

    Abs. The most comprehensive SMC article published to date. The problems of an economical preparation of resin mix and an efficient glass impregnation are discussed. SMC impregnator types are described and illustrated.

318. Signorelli, E., "Moulds For SMC and BMC," *Poliplasti*, Oct. 1974.

Abs. Molds for compression, transfer, and injection molding are considered.

319. Harris, B., and Cawthorne, D., "Strength and Toughness of SMC," *Plas. Poly.*, Oct. 1974, pp. 209–216.

Abs. Mechanical properties and fracture toughness of SMC and DMC were measured and reported.

320. Meyer, R. W., McGranahan, F., and Evans, D., U.S. Patent 3,847,707, "Laminating Apparatus Having Dual Doctor Blades," Nov. 12, 1974.

Abs. This patent describes the equipment required for incorporating continuous rovings in an SMC sheet.

321. Mandy, F., "Market and Applications of SMC: New Developments in LP Formulations," Preprints 8th RP Congress, Brighton, England, pp. 33–39, Nov. 11–14, 1974.

Abs. The SMC market and applications in Europe, Japan, and in the U.S. are documented. The results of a design study of rib constructions are also presented.

322. Methven, J. M., "Requirements for Fire Retardant Polyester Resins in the Future," Ibid., paper no. 15, pp. 107–110.

Abs. Current requirements of fire testing and the relevance of such tests are discussed.

323. Deis, W. H., and Ness, W. W., "Magnesium Oxide and Hydroxide for SMC," *Mod. Plas.*, Nov. 1974, pp. 94–98.

Abs. Magnesium hydroxide affords a slow to moderate rate of thickening, while high-surface activity magnesium oxide is a faster-acting thickener. A condensation of Item 288 above.

324. Anon., "Here's How to Repair Molded SMC Parts on the Moving Assembly Line," *Plas. Des. & Proc.*, Nov. 1974, pp. 21–23.

Abs. Work procedures are detailed for repairing gouges, cracks, pits, and edge chips on SMC parts already installed on automobiles.

325. Dastur, S. M., "Mould and Part Design for GRP Moulding Compound," *Op. Plas.*, Dec. 1974, pp. 35–38.

Abs. Design principles to obtain the best molding behavior from DMC are listed.

326. Jain, M. K., "Introduction to DMC and SMC," *Op. Plas.*, Dec. 1974, pp. 39–42.

    Abs. Applications of DMC and SMC are discussed. Tool design and processing notes are included.

327. Pullman, G. A., "Finding an Optimum Resin/Glass Combination for SMC," *Plas. Des. & Proc.*, Dec. 1974, pp. 17–18.

    Abs. The author discusses strand length, filament count, and filament diameter, as they affect physical properties in an SMC system.

328. Anon., "High Performance Composites Replace Metals," published by Morgan Grampian, London, 1974, pp. 104–105.

    Abs. Tables of physical properties of BMC and other materials are given.

329. Gaylord, M. W., *Reinforced Plastics*, 2nd edition, Cahners Pub. Co., Boston, 1974.

    Abs. A brief description of BMC and SMC is provided as well as molding notes and a discussion of raw materials.

1975

330. Wilson, E. L., McCluskey, E. J., Bell, R. Z., and Linak, R. T., U.S. Patent 3,861,982, "Apparatus for Producing SMC," Jan. 21, 1975.

    Abs. A horizontal belt-type SMC machine is described together with the resin-dispensing system.

331. Fry, D. P., U.S. Patent 3,862,064, "Molding Compositions Containing Unsaturated and Saturated Polyesters, Cellulose Esters and Monomeric Material," Jan. 21, 1975.

    Abs. An unsaturated polyester resin composition for SMC comprising an unsaturated polyester, a copolymerixable monomer, a saturated polyester liquid obtained by esterifying a polyhydric alcohol with a polybasic acid, and a cellulose ester of a fatty acid of up to 6 carbon atoms.

Annotated Bibliography 301

332. Shenk, W., "Increasing Automotive Markets for GRP," *Euro. Plas. News*, Jan. 1975, p. 37.
    Abs. Statistics of FRP use in automotive markets in the U.S. and the U.K. are given.

333. Waldrek, J. W., "Internal Mould Release Agents in FRP Processing," *Plas. Des. & Proc.*, Jan. 1975, pp. 23–25.
    Abs. Internal lubricants for FRP molding compounds are discussed.

334. Bauer, S. H., "BMC's Compatable With Runnerless Injection Molding," Ibid., pp. 8–11.
    Abs. The concept of injection molding SMC in a runnerless mold is described. Sketches are provided of a conventional 3-plate mold and of a runnerless mold.

335. Uffner, M. W., "Effects of Compounding on the Performance of PEP Promoted SMC," Proceedings SPI RP/CI 30th Ann. Meeting, Washington, D.C., Feb. 4–7, 1975, Sec. 1-A.
    Abs. PEP organometallic promoters reduce the cure times of SMC 15 to 60% without creating pregel problems. Shelf life data are presented involving several thickeners and inhibitors.

336. Sundstrom, C. A., Collister, J., and Hays, C. W., "Low Density SMC and BMC," Ibid., Sec. 1-B.
    Abs. SMC and BMC compounds containing hollow glass spheres are discussed. Densities in the range of 1.2 to 1.4 g/cc are reported.

337. Gruenwald, R., and Walter, O., "Fifteen Years' Experience With SMC," Ibid., Sec. 1-C.
    Abs. A brief history of SMC technology development and an illustrated description of impregnator design evolution is given. Early applications are mentioned and data on outdoor weathering are included.

338. Calendine, R., and Blount, W. W., "Polyester Resin Based on TMPD Glycol for SMC Applications," Ibid., Sec. 1-D.
    Abs. A new polyester resin for SMC based on TMPD and ethylene glycol is described. It is claimed that the new resin in SMC can be pigmented and has hydrolytic stability.

339. Riew, C. K., Rowe, E. H., Backderf, R., and Guiley, C., "A New Toughener Designed for BMC," Ibid., Sec. 1-E.

Abs. Abstract only in proceedings. Paper was passed out at the meeting. It is claimed that impact strength and crack resistance can be improved in BMC and SMC formulations by adding a CTBN material adsorbed on calcium carbonate.

340. Kubota, H., "Curing of Highly Reactive Polyester Resin under Pressure," Ibid., Sec. 1-F.

Abs. The kinetics of polyester cure under pressure are investigated.

341. Fekete, F., "Maximized Specific LS-FR Resin and SMC Compositions," Ibid., Sec. 1-G.

Abs. This paper deals with methods for providing maximum fire retardance to polyester resins using new or modified synthesis procedures to control the degree of functionality, reactivity, structural configuration, and molecular weight.

342. Fletcher, C. W., and Waldeck, J. W., "Parameters Affecting the Selection of Metallic Soaps for Polyester Molding," Ibid., Sec. 8-A.

Abs. The type and grades of internal lubricants are reviewed. Viscosity effects with MgO and $Mg(OH)_2$ are presented.

343. Grotke, M., and Garst, J., "Direct Digital Computer Control of Polyester Batch Manufacturing Process," Ibid., Sec. 10-A.

Abs. The development of a direct digital computer control system for the production of polyester resins is described.

344. Schmitterger, P. A., "Thermosetting Polyester Molding Compounds," Ibid., Sec. 10-B.

Abs. A new line of Reichhold dry granular BMC's is described.

345. Fackler, M., and Rudy, A. J., "Designs and Tooling of a Unique Vehicle in SMC," Ibid., Sec. 11-A.

Abs. The history of the development of the Kawacki Jet Ski, a new water vehicle molded of SMC, is related.

346. Bauer, S. H., and Meister, G., "Runnerless Injection Molding of BMC Polyester Compounds," Proceedings SPI RP/CI 30th Ann. Meeting, Washington, D.C., Feb. 4–7, 1975, Sec. 11-A.

Abs. The use of cold manifold runnerless molding of BMC parts is described. Sketches of such molds are included.

347. Mackrael, T., "A New Injection Molding System for BMC Processing," Ibid., Sec. 15-C.

Abs. Injection molding of BMC provides advantages over transfer or compression molding. Sketches relate the effect of injection pressure on physical properties.

348. Callahan, M. L., "Selecting Reinforced Thermoset Polyester for Injection Molding," Ibid., Sec. 20-B.

Abs. A very sketchy discussion of the materials used in injection molding of BMC is presented.

349. Ward, D. D. and Robbins, G., "GRP and Its Impact on Business Machines," Ibid., Sec. 20-D.

Abs. The requirements of business machine housings are described and the processes of injection molding, transfer molding, and compression molding are reviewed.

350. Deller, T. E., Injection Molding Glass Reinforced Polyesters for Electrical Applications," Ibid., Sec. 20-E.

Abs. The development of the BMC injection molding process has resulted in a totally automated process for producing electrical parts. A number of applications are cited and illustrated.

351. Parr, A. J., and Methven, L. M., "Investigations in Automated Process Control of the Mixing of DMC's," Ibid., Sec. 21-A.

Abs. This paper relates the measurement of power consumption of a DMC mixer with physical properties of the compound produced and suggests ways of improving the compounds.

352. Bolen, G. N., "Selecting Glass Reinforcements," *Plas. World*, Feb. 17, 1975, pp. 32–35.

Abs. A rundown of the OCF products available for reinforcing FRP parts is provided.

353. "Production Testing of SMC Content for Repeatability," *Plas. Des. & Proc.*, Feb. 1975, pp. 19–20.

   Abs. Quality control procedures for measuring initial paste viscosity, glass content, and square foot weight are described.

354. Bradley, I. G., "Other Production Methods-DMC's and Moulding Compounds," BPF, RP in Electricals Symposium, Brighton, England, Feb. 1975, paper No. 6.

   Abs. The manufacture of DMC and SMC is covered. Molding techniques are reviewed, and physical properties are considered.

355. Tortulano, F. W., "Try RP for Close Tolerance High Volume Parts," *Plas. World*, May 1975, pp. 39–42.

   Abs. SMC and BMC parts are shown for the Xerox copier.

356. "Improved Polyester Molding Materials," *Poly. Age*, May 1975, pp. 136–197.

   Abs. DMC and SMC compounds are discussed in relation to fire retardancy, low shrink, and pigmentability.

357. "RP-Materials, Moulds, Finished Products," *Plas. Mat. Design*, May 1975, pp. 29–34.

   Abs. Suppliers of raw materials are listed, and their products are described.

358. "Versatile Mixing Unit Allows Variable Intensity," *Plastics Technology*, May 1975, pp. 11–12.

359. Comey, J. F., "Injection Molding FRP—Now More Adaptable Than Ever," *Plas. World*, Aug. 1975, pp. 55–58.

   Abs. A number of injection molded polyester parts are shown and discussed that are made in Glastic's Jefferson, OH, and Chicago, IL, plants. Cost comparisons with aluminum are provided.

360. Tortolano, F. W., "High Strength FRP Outperforms Steel," *Plas. World*, Nov. 1975, pp. 46–48.

   Abs. PPG's HMC and XMC are discussed.

361. Lottig, R., Neiman, T., and Stevenson, D., "Pigmented LP Polyester Systems," SPE NATEC, "Plastics in Appliances," Louisville, KY, Nov. 1975, pp. 214–215.

362. Nendal, C. G., "Past, Present and Future of FRP in Appliances," Ibid., pp. 161–165.
363. "Redesign with FRP Brightens Circuit Breaker Economics," *Plas. World*, Dec. 22, 1975, p. 45.

    Abs. Rostone BMC molded circuit breakers are discussed.
364. Schneberger, G. L., *Understanding Paint and Paint Processes*, Hitchock Pub. Co., Wheaton, IL, 1975.

    Abs. An excellent text for shop personnel. Covers: paint concepts, paint components, paint properties, types of coatings, surface preparation, phosphating, application equipment, paint film defects, repair procedures, testing paints, etc.

1976

365. Waldeck, T. W., "Selecting Mold Release Agents for FRP," *Plas. World*, Jan. 9, 1976, pp. 40–42.

    Abs. Uses of metallic stearates in BMC and SMC are given for metallic stearate type and concentration for various FRP products.
366. Thomas, R. R., "Business Machines: Cost Savings Applications in SMC," Proceedings 31st SPI RP/CI ann. meeting, Washington, D.C., Sec. 1-C.

    Abs. The requirements of business machine housings are presented. Several applications are discussed and illustrated.
367. Cross, J., and Nenadal, C. G., "Appliances Designed with FRP," Ibid., Sec. 1-E.

    Abs. Room air conditioner parts are discussed.
368. Shreve, J. T., and Tropp, F. E., "Resilient Polyester System for FRP," Ibid., Sec. 2-A.

    Abs. The Koplac™ 3102-5 SMC resin system is discussed.
369. Maxel, J. M., "Epoxy-Steel Tooling for the SMC Process," Ibid., Sec. 2-B.

Abs. This paper covers the use of low-cost prototype tooling coupled with low-pressure SMC.

370. Boyd, J., "Observation on the Mechanism of Sink in SMC Molding," Ibid., Sec. 2-C.

Abs. A study was made of the effect of resin composition, thermoplastic additive, thickening agent, catalyst, filler, and reinforcement on sinks in SMC pastes.

371. Williams, G. L., "A New Approach to SMC," Ibid., Sec. 2-D.

Abs. SMC II (OCF's low-pressure SMC) is presented including advantages, formulations, molding parameters, recommendations for presses and molds, testing, and discussion of several market requirements.

372. Atkins, K. E., Koleske, J. U., Smith, P. L., Water, E. R., and Mathories, V. E., Ibid., Sec. 2-E.

Abs. Overall shrinkage of a molding compound during the polymerization reaction is prevented by thermal expansion of the thermoplastic additive. The differences in performance of thermoplastics is due to their thermal coefficient of expansion, glass transition temperature, and polarity.

373. Fletcher, C. W., "The Use of Magnesium Stearate in RP Unsaturated Polyester Molding Composites," Ibid., Sec. 4-A.

Abs. The object of this paper is to examine the characteristics of magnesium stearate as compared with those of calcium and zinc stearates.

374. Cooper, T. I., and McKay, M. S., "Method of Introducing Additives to Molding Compound Formulations," Ibid., Sec. 4-D.

Abs. The use of a soluble film package for adding mold release, catalyst, pigments, etc. to a formulation is described.

375. Vockel, R. L., "Pigmentation Techniques for High Glass Deep Color SMC," Ibid., Sec. 7-B.

Abs. The techniques used to develop molded-in color SMC are outlined. A troubleshooting guide is included.

376. Maaghul, J., and Potkanowicz, E., "HMC™—A High Performance SMC," Ibid., Sec. 7-C.

Abs. Initial SPI paper on HMC. Includes formulations, properties, applications.

377. Bradish, F. W., "RFI/EMI Shielding for SMC and BMC," Ibid., Sec. 7-D.

    Abs. The use of shields to minimize RFI and EMI on SMC and BMC molded parts is discussed.

378. Methven, J. M., "Weibull Distributions in the Interpretation of the Mechanical Properties of Polyester Molding Compounds," Ibid., Sec. 7-E.

    Abs. This paper discusses the interpretation of physical property data in terms of Weibull distribution functions.

379. Keating, J. Z., "Filler Management in Polyester Resin Systems," Ibid., Sec. 8-C.

    Abs. This paper outlines the techniques of selecting and using fillers, provides information on particle size analysis, handling fillers, mixing fillers, effect of fillers on physical properties and flammability, filler quality control, and an economical analysis of filled systems.

380. Greenzweig, J. E., and Pickering, T. L., "Flow Properties of Calcium Carbonate Filled Polyester Resins," Ibid., Sec. 8-D.

    Abs. The influence of particle size, particle size distribution, and filler loading on the rheological properties of molding compounds is presented.

381. Grisch, W. E., "Cost Reduction Corrosion Resistant Structures Using SMC," Ibid., Sec. 9-F.

    Abs. This paper discusses adapting automotive SMC technology to corrosion structures for cost reduction and quality improvement.

382. Colombo, D. E., "Injection Molding—A Market Evaluation for Thermoset Polyesters," Ibid., Sec. 10-D.

    Abs. This paper describes the market for injection-molded BMC in automotive, small appliances, and electrical applications.

383. Ward, D. D., "Injection Molding of FRP for Electrical Applications," Ibid., Sec. 10-E.

    Abs. The case history of an electrical device development is presented with the advantages and disadvantages of injection-molded BMC.

384. Ross, P. M., "Development of a Light Weight Plastic Hood Panel for High Volume Passenger Car Applications," Ibid., Sec. 11-A.

Abs. A two-piece adhesively bonded hood utilizing a thin SMC outer panel is described. Weight savings is the chief advantage.

385. Austin, V., "Responsibilities of the Coatings Supplier and Molder to the Automotive Industry," Ibid., Sec. 11-B.

Abs. Random efforts have been made with very little benefit to upgrade the process in the coating of plastic automotive parts, particularily SMC. The industry itself regards the priming or painting operation as a necessary evil, not admitting that it would not have the molding job in the first place without the painting capability. Recognition of the responsibility of the coating manufacturer to both the molder and the automotive industry has prompted the major part companies to step forward as the leaders in a concentrated effort to improve the finishing process at the molder and follow through with the automotive assembly plant.

386. Bauer, S. H., "Injection Molding Thermosetting BMC," Ibid., Sec. 13-A.

Abs. Injection molding of BMC has resulted in cost savings through reduced cycle times. Runnerless molding saves material waste.

387. Levy, M., "Reinforced Granular Polyesters," Ibid., Sec. 14-E.

Abs. No paper in proceedings. Paper obtained from author. A solid polyester resin for use in compounding granular injection molding compounds is described.

388. Davies, J. W., "SMC Answers to the Automotive Challenge," Ibid., Sec. 16-A.

Abs. This paper outlines some of the steps taken to overcome brittleness, make repair simpler, lower weight, etc. for automotive applications.

389. Ackley, R. H., "XMC™—Structural FGRP for Matched-Metal-Die Molding," Ibid., Sec. 16-C.

Abs. Initial SPI paper on XMC. Included are properties and photos of proposed applications.

390. Todd, W. H., "Compression Molding Equipment Development," Ibid., Sec. 18-B.

Abs. The GM programmable Force/Velocity Control system provides resistance to the press ram at the four press corners during the final 1-inch press travel and during the initial press opening, maintaining platen parallelism.

391. Charters, C. A., "Getting the Most from an SMC Part," *Plas. Engr.*, Feb. 1976, p. 8.

Abs. Some SMC design principles are reviewed.

392. Shenk, W., "GRP In Automobiles," *Poly. Paint Col. Jou.*, Mar. 24, 1976, pp. 261–263.

Abs. The market in the U.S. and Europe for SMC and DMC parts in automotive applications is reviewed.

393. "SMC Industry Begins to Sell Itself," *Reinf. Plas.*, Apr. 1976, pp. 96–97.

Abs. Six papers were given by the Polyester Compound Group of BPF to discuss SMC and DMC.

394. "Designing Successfully with GRP," *Plas. Des. Forum*, May 6, 1976, pp. 54–61.

Abs. A hasty description of glass fiber products is given followed by a list of four basic GRP design principles and the 10 stages of design from concept to production.

395. Williams, G. L., "Lower Molding Pressures Brighten Outlook for Shorter-Run SMC Parts," *Plas. World*, May 17, 1976, pp. 47–50.

Abs. OCF's SMC II is described. See Item 371 above.

396. Worldwide Plas. Develop. Ltd., "Extend Your Range With Steel Inserts," *Brit. Plas. Rubber*, May 1976, pp. 37–38.

Abs. Details are given for the Aro process of injection molding and compression molding around steel structural forms.

397. "FRP Design for Heavy Duty Electricals," *Plas. World*, May 22, 1976, pp. 39–41.

Abs. New applications for BMC in circuit breakers and electrical switchgear are shown.

398. Corneliussen, L., "Moulding of GRP," *Plastnytt*, June 1976, pp. 19–23.

Abs. Compression molding of GRP involving SMC, DMC, cold pressing, and resin-inject molding are discussed.

399. OCF, "Sheet Molding Compound," OCF publication 5-TM-6991-A (June 1976), pp. 45–42, "Techniques of Finishing SMC."

400. "Freeman Chemicals SMC, DMC and You," Freeman Chemicals Ltd., Ellesmere Port, U.K.

Abs. Information is outlined on market trends, applications, part and tool design, molding notes, etc.

401. "The Fundamentals of Mold Design—Key Element in Tooling for FRP Molding," *Plas. Design & Processing*, Aug. 1976, pp. 13–14.

Abs. This article discusses mold materials, mold components, general design rules, etc.

402. Hays, C. W., "FRP—New Materials and Applications in SMC and BMC," SPE Western Section, Seattle, WA, pp. 1–11.

Abs. The dimensional stability, strength, machining, and paintability are discussed.

403. "Basic Press Types for FRP Molding," *Plas. Design & Processing*, Sept. 1976, pp. 18–19.

Abs. The type of press, performance criteria, and requirements for several FRP processes are discussed.

404. Penfold, R., "Progress in Moulding Pieces of Large Diameter in GB," *Mat. Plas. Elast.*, Sept. 1976, pp. 701–706 (in Italian).

Abs. This study included processes and machinery for injection molding DMC

405. Wolf, G. M., "Painted Elastomers in Tomorrow's Automobile," SAE Meeting, Dearborn, MI Oct. 18–24, 1976, SAE paper 760732.

Abs. Painted and color-impregnated elastomerics are used on exterior body components. Flexible urethanes, microcellular, and injection-molded ethylene propylene terpolymer, thermoplastic polyolefins, all painted matching or contrasting body colors, color impregnated or painted vinyl and color impregnated ethylene ethyl acrylate are being utilized for applications such as bumpers, stone deflectors, front ends, rear quarter extensions, and body side moldings. The initial incentive was to reduce damageability. Lately government regulations, both state and national, dictated energy-absorbing systems, and more recently the energy crisis has prompted usage in some applications to reduce bodyweight (author).

406. Burns, R., and Gandhi, K. S., "The Rheology of Glass Fibre Reinforced Moulding Compounds," Preprints BPF RP Group Meeting, Brighton, England, paper No. 4, pp. 63–70.

    Abs. The strength of a DMC molding was shown to depend on fiber orientation, which depends on the rheological properties of the compound. There is a correlation between fiber orientation and the power law-index. Additions of glass fiber with suitable sizings makes DMC more viscous and shear thinning than the addition of insoluble coated fibers.

407. Dawson, T., "Polyester Moulding Compounds—Comparison of In-House Manufacture with Packaged," Ibid., paper No. 5, pp. 71–76.

    Abs. The advantages of in-house compounding are cost savings and more flexibility in production planning. The disadvantages are that the molder takes on full responsibility for the quality of the compound and a larger technical staff is required.

408. Garholm, K., and Johansson, E., "Optimizing Economical Benefits by Design and Integration of SMC Parts for Vehicle Bodies," Ibid., paper No. 11.

    Abs. An economical analysis is presented vs. steel for several SMC parts on Swedish cars.

409. "GRP Technology and Practice," BTR RP Ltd., Uxbridge.

    Abs. A discussion is given of the design support, quality control, and production facilities at BTR RP Ltd.

410. *Unsaturated Polyester Technology,* Seminar papers, Gordon & Breach Science Pub. Inc., New York, 1976.

    Abs. The papers in this book indicate how new developments may be made in higher chemical resistance, improved flame retardancy, improved SMC and BMC materials, etc.

411. Rapp, R.S., "High Strength Molding Compounds," presented at the SPI RP/CI RTPMC meeting, Oct. 1976.

412. Deller, T. E., "Injection Molding of Thermoset Glass Polyesters," *Insulation Circuits*, Nov. 1976, pp. 68–71.

   Abs. Injection molding is discussed with particular reference to cost comparisons with compression and transfer molding of electrical parts.

413. "Polyester Compounds: Improving Technology Spurs Growth," *Eur. Plas. News*, Nov. 1976, pp. 19–22.

   Abs. Developments in the preparation of SMC and DMC and new types of SMC available from PPG Industries are discussed.

414. Atkins, K. E., Gentry, R. R., Berger, S. E., and Schwarz, E. G., "New Silane Agent for ATH," *Plas. Engr.*, Dec. 1976, pp. 23–24.

   Abs. Tests are described that demonstrated that the physical properties of BMC are improved by pretreating the ATH filler with a silane agent.

415. Methven, J., "Flow Behavior of SMC and DMC," *Plas. Rubb.*, Dec. 1976, pp. 149–157.

   Abs. The flow and cure characteristics were determined by instrumented spiral flow and platen movement tests.

416. "Thermosetting Compounds," *Engr. Plas. Guide*, published by *European Plas. News*, 1976.

   Abs. Details are given for many compounds including SMC and DMC.

1977

417. Kobayashi, G. S., and Pelton, R., "Process Development for Fabricating Sculptured Interior Aircraft Panels Using SMC," Proceedings SPI 32nd RP/CI Annual Meeting, Washington, D.C., Feb. 8–11, 1977, Sec. 2-A.

   Abs. LP SMC was used for interior panels for the redesigned Boeing 747. Including attachment bosses and stiffening ribs was an important criteria for SMC selection. Profusely illustrated.

418. Griffith, R. M., Shanoski, H., and van Essen, W., "In-Mold Coating of SMC Moldings," Ibid., Sec. 2-C.

Abs. The General Tire in-mold coating development is described.

419. Maxel, J. M., "Rheology of SMC at Molding Temperature," Ibid., Sec. 2-E.

Abs. The use of Rheometrics mechanical spectrometer to study SMC flow is discussed.

420. Massey, F. L., "Boss and Fastener Design Specifications," Ibid., Sec. 2-F.

Abs. A study of the effect of boss type, hole type, hole diameter, and glass content on driving torque, torque failure, and pull-out force was made on 3 types of screws. Recommendations are given for boss size and hole diameter for several sizes of self-tapping screws.

421. Hays, C. W., "SMC Replacement of Foamed Engineering Thermoplastics: A Case History of a Computer Printer Cover," Ibid., Sec. 2-G.

Abs. The step-by-step development of a computer printer cover from RIM to foamed polycarbonate to SMC is outlined.

422. Moulson, T. J., and Mathur, K. K., "Low Viscosity Limestone Fillers for SMC," Ibid., Sec. 3-F.

Abs. Low-viscosity calcium carbonate fillers were prepared by surface treatment and modification of particle size distribution.

423. Pluddemann, E. P., and Stark, C. L., "Surface Modification of Fillers and Reinforcement in Plastics," Ibid., SEc. 4-C.

Abs. A study was made of silane modification of mineral surfaces to improve adhesion, water resistance, filler wet out, and viscosity control.

424. Atkins, K. E., Gentry, R. R., Gandy, R. C., Berger, S. E., and Schwartz, E. G., "Silane Treated ATH: A New Formulating Tool for FR Polyester FRP," Ibid., Sec. 4-D.

Abs. A silane treated ATH that permits increased filler loadings, reduced wet-out time, and improved mold flow and glass dispersion is discussed.

425. Monte, S. J., Sugarman, G., and Seeman, D. J., "Titanate Coupling Agents—Update 1977," Ibid., Sec. 4-E.

Abs. The functions of titanate coupling agents and the factors necessary for a successful application are discussed.

426. Grish, W., and Maxel, J., "Performance of SMC in Severe Environments," Ibid., Sec. 5-A.

Abs. The properties of several corrosion resistant SMCs are listed.

427. Shaw, A. W., and Schortenberger, J., "The Electrostatic Application of Thermosetting Materials," Ibid., Sec. 6-A.

428. Fekete, F., Howatineck, F., and Fazekas, E., "A Study of the Influence of Specific Variables on Viscosity Relationships and Surface Properties of SMC Formulations Based on Special Calcium Carbonate Fillers," Ibid., Sec. 7-B.

Abs. Four categories of calcium carbonates were compounded into pastes and SMCs and the effects on viscosity were studied.

429. Parker, F. J., "The Benefits of Injection Molding of DMC and the Effect on Strength," Ibid., Sec. 6-F.

Abs. Injection molding of DMC has faster production rates and easier handling, which reduces costs at a slight sacrifice in physical properties.

430. Burns, R., and Gandhi, K. S., "The Rheology of Glass Reinforced Polyester Molding Compounds," Ibid., Sec. 7-C.

Abs. The strength of FRP moldings is shown to depend strongly on the fiber orientation, which in turn depends on the rheological properties of the DMC.

431. Atkins, K. E., Gentry, R. R., Mathews, V. E., and Gandy, R. C., "The Pigmentation and Shrinkage Control in Fiber Reinforced Molding Compounds," Ibid., Sec. 7-D.

Abs. A study of the effects of various polyester resins with several thermoplastic additives is reported. The effects of shrinkage and pigmentation have been stressed in this study.

432. Smith, R. J., and Jutte, R. B., "The Source and Nature

of Mechanical Property Variability in SMC," Ibid., Sec. 7-E.

Abs. The influence of flow and the directionality of the glass reinforcement in SMC have been identified in this work as contributing to the variability in SMC properties.

433. Ampthor, F. J., Marple, R. K., and Lowry, J. L., "Chemical Thickening Test for Polyester Resins," Ibid., Sec. 7-G.

Abs. A resin-only chemical-thickening test is described.

434. Olsen, N., Powell, R., and Burer, E., "The Use of SMC in Aerodynamic Deflector," Ibid., Sec. 12-A.

Abs. A truck cab wind deflector is discussed.

435. Delaney, D. E., "Design of an Experiment to Solve Porosity on Polyester Panels," Ibid., Sec. 12-D.

Abs. The result of a task force investigation into the causes of porosity in 1976 Grand Prix front ends is presented.

436. Hellner, G., "Optimizing Economical Benefits by Design and Integration of SMC Parts for Vehicle Bodies," Ibid., Sec. 12-E.

Abs. The advantages and disadvantages of SMC in automobile parts is presented. Case histories of 6 parts are documented.

437. Davies, J. W., "Transportation Market vs. Industry Capacity," Ibid., Sec. 12-F.

Abs. The SMC transportation market is discussed with projections for future growth.

438. Lundberg, J. A., and Coats, C. E., "Energy Absorbing Polyesters for Low Profile SMC," Ibid., Sec. 16-A.

Abs. Ashland's tougher SMC system is discussed.

439. Collister, J., and Powell, R., "Premi-Glas™ Energy Absorbing SMC," Ibid., Sec. 16-B.

Abs. SMCs with improved resistance to brittle failure are discussed.

440. McGarry, F. L., Korve, E. H., and Riew, C. K., "Improving Crack Resistance of BMC and SMC," Ibid., Sec. 16-C.

Abs. Microcracking in laminates can be reduced by including liquid rubbers in the polyester matrix.

441. South, A., and Workman, R. T., "Rubber Modifiers for Thermoset Polyesters," Ibid., Sec. 16-D.

    Abs. Solprene™ modified butadiene-styrene rubbers as impact modifiers for SMC and BMC are described.

442. Marker, L., and Ford, B., "Rheology and Molding Characteristics of Glass Fiber Reinforced SMC," Ibid., Sec. 16-E.

    Abs. An instrumented mold for the study of pressure and temperature effects on compounds during cure is described. The rheological properties of the compounds used were studied with a capillary rheometer and a Rheometrics mechanical spectrometer.

443. Thomas, R. R., "A Case History of the Use of SMC in a Highly Dimensional and Creepfree Application," Ibid., Sec. 16-B.

    Abs. The use of SMC in a microfilm reader-printer is described.

444. Gardner, I. J., "Elastomeric Toughening of Thermoset Resins," Ibid., Sec. 17-C.

    Abs. No paper in the proceedings.

445. Bassford, F. E., "Low Profile Polyester Resins," *Plas. Rubb. Int.*, Mar.–Apr. 1977, pp. 81–82.

    Abs. Polyester resins are discussed in detail with reference to DMC and SMC.

446. Union Carbide, "No Painting for Glass/Polyester Molded Parts," *Mach. Des.*, Mar. 24, 1977, p. 12.

    Abs. A new low-profile additive for polyester compounds provides pigmentation with zero shrinkage.

447. Clavadetcher, D. J., and Bradish, F. W., "Competition for New Applications Pushes Improvements in SMC and BMC," *Plas. Des. & Proc.*, Mar. 1977, pp. 43–48.

    Abs. Comments are provided on low-density and energy absorbing SMC and BMC.

448. Swaneveld, I. A., "The European SMC Market and the Opportunities for XMC/MHCR in Particular," Preprints, International Symposium on Polyester Moulding Compounds, Geneva, May 5–6, 1977, Paper No. 3.

449. Bowyer, E. G., "Recent Advances in Technology of Polyester Resins and Their Use in Moulding Compounds," Ibid., Paper No. 5.

450. Hentschel, H., and Lottler, C., "Flammwidrige Polyesterformassen mit Aluaminumhydroxid," Ibid., Paper No. 8.
451. Harms, W., "Neue Entwicklungen von riesel fahlgen Polyesterpressmassen," Ibid., Paper No. 9.
452. Luchini, A., "SMC Design and Moulding Criteria," Ibid., Paper No. 10.
453. Butler, M. J., "Moulds and Machinery," Ibid., Paper No. 11.
454. Seamark, M. J., "Development of ERF Vehicle Cab," Ibid., Paper No. 13.
455. Chiesi, F., "The Development of Parts with High-Dimensional Accuracy," Ibid., Paper No. 15.
456. Houlton, P. R., Izzard, K. T., and Newton, G. P., "Selection of Catalysts for Dough Moulding Compounds," Ibid., Paper No. 6.
457. Burns, R., "Effect of Fibre Parameters on the Properties of Glass Fibre Reinforced Polyester Moulding Compounds," Ibid., Paper No. 7.
458. Hodgson, W. D., "Some Interesting Applications for Polyester Moulding Compounds," Ibid., Paper No. 16.
459. Van Gasse, R. L., "Application of BMC in the Building Industry," Ibid., Paper No. 17.
460. Hodgson, W. D., "Why Use SMC and DMC?" Plas. & Rubb. Inst. Seminar, "1980's Car—What Future for Plastics," May 1977, Paper No. 8.

    Abs. It is predicted that DMC and SMC parts will play an increasingly important role in automobiles.
461. Kent, G., "Here's how to Pretreat Parts for Painting," *Plastics Technology* (June 1977), pp. 69–72.

    Abs. This article discusses the various surface contaminations on a plastic part, their effect on the coated part if not properly removed, and the recommended cleaning cycles for removal using a power washer.
462. Englehart, W. H., and Richter, C. S., "Boss-Fastener Designs for SMC," *Mod. Plas.*, July 1977, pp. 72–77.
463. Grove, S. A., "Engineering Mixing Systems for Filled Polyester Resins," *Plas. Des. & Proc.*, Jul. 1977, pp. 55–63.

Abs. Shear requirements, blade type, and filler loading are discussed.

464. Pennington, D., and Shortall, J. B., "Injection Moulding DMC/SMC," Interplas 77 Conference, "Optimizing the Use of Resources," Birmingham, England, Paper No. 5.

    Abs. The advantages of injection molding thermosets is discussed.

465. Moeseller, H. J., "Use of Polyester SMC and DMC in European Vehicle Construction," *Ambeits. Verst. Kunst.*, Oct. 1977, Paper No. 35 (in German).

    Abs. A review is provided of the use of FRP parts in vehicle construction.

466. Clavadetcher, D. J., "Progress in SMC/BMC Processing," *Arbeits. Verst. Kunst.*, Oct. 1977, Paper No. 8.

    Abs. Data on Premi-Glas™ materials are provided.

467. Ellerhorst, Jr., H., "Coating Technology Trends: A Ten-Year Forecast," *Modern Paint & Coatings*, Oct. 1977, pp. 101–104.

    Abs. As a result of a recent Delphi study, this article takes a good look at the prospects for chemical coatings systems in the various markets for the decade ahead.

468. Parker, F. J., "Injection Molding of BMC With Minimum Strength Loss," *Mod. Plas.*, Dec. 1977, pp. 54–58.

    Abs. Based on Item 429 above.

1978

469. Hoefele, A. J., "FRP Finds a New Home in the Electrical Industry," Proceedings SPI 33rd RP/CI Ann. Mtg., Washington, D.C., Feb. 7–10, 1978, Sec. 3-E.

    Abs. A new flame-retardant low smoke BMC was injection molded to provide low-cost electrical connectors.

470. Bradish, F. W., "Conductive SMC/BMC Composites for RFI/EMI Shielding," Ibid., Sec. 4-A.

    Abs. The development of conductive compounds for use as electromagnetic shields is described.

471. Izzard, K. J., and Newton, G. P., "The Dependence of Degree of Cure and Shelf Life of DMC with Respect to Catalyst Selection," Ibid., Sec. 5-C.

Abs. It was shown in this study that the level of free styrene in a molding may be used as a guide to the degree of cure and free benzaldehyde content may be used to predict the further useful life of the DMC. TBPO is the most useful catalyst at low melt temperatures (120–130°C), but for LP compounds TBPB or the perketals are preferred.

472. Ampthor, F. J., Huminski, F. M., Lowrey, J. R., Marple, R. K., and Trimble, S. M., "Chemical Thickening and Maturation Control for Production SMC" Ibid., Sec. 9-E.

Abs. The use of Paraplex™ CM-201 maturation control agent is described.

473. Liebold, R. W., "An Optimum Design in Mass Production of SMC Bumpers Suitable to Being Lacquered for European Cars," Ibid., Sec. 13-A.

Abs. The development of the Porsch 924 bumpers is discussed. Mold design, molding conditions, materials comparisons, test data accumulated, quality control, are all presented. Temperature and pressure transducers were placed in the mold to monitor molding conditions. No painting data are included. Everything is aimed at improving the molded part.

474. Sherman, E. A., "Heat Transfer Phenomena in Compression Molding Glass Fiber Reinforced SMC," Ibid., Sec. 14-F.

Abs. A study of heat flow in a compression mold was investigated. It was found that: a. Differences in adjacent part features will disturb the thermal equilibrium of the mold sufficiently to alter the curing process, b. Heat transfer features should be engineered into the mold, c. Heat flow engineering in the mold can reduce cycle times.

475. South, A., Dix, J. S., and Hill, H. W., "A Rubber Modifier to Improve Toughness and Pigmentability of SMC and BMC," Ibid., Sec. 19-A.

Abs. Abstract only in the proceedings. Copy passed out at meeting. A carboxy-terminated SB block polyer (Solprene™ 312) is described. It is claimed that energy-absorbing properties are imparted to BMC/SMC as well as low shrinkage, no sink marks, and good pigmentability when this modifier is used. Use of a cure promoter further reduces shrinkage and provides higher gloss.

476. Rowe, E. H., and Howard, F. H., "A New Toughening Additive for BMC and SMC," Ibid., Sec. 19-B.

Abs. A hydroxy terminated polyether as a toughening additive in BMC is described.

477. Gardner, I. J., and Fusco, J. V., "High Impact Resistant Polyester Composites," Ibid., Sec. 19-C.

Abs. No paper in proceedings.

478. Marple, R. K., Ampthor, F. J., and Trimble, S. M., "Effect of Resin and Syrup Variations on SMC/BMC Production Variables," Ibid., Sec. 19-D.

Abs. A study was made of the effect of physical constant variables of Paraplex™ P-340 and P-701 syrup on compound properties. The polyester had more influence than did the syrup.

479. Owen, M. J., Thomas, D. H., and Found, M. S., "Flow, Fibre Orientation, and Mechanical Property Relationship in Polyester DMC," Ibid., Sec. 20-B.

Abs. DMC flow in a mold leads to predictable fiber orientation patterns, which can be related to mechanical properties.

480. Just, H. A., and Englehart, W. H., "Low Pressure SMC-Approaches for Tooling," Ibid., Sec. 21-D.

Abs. The use of low-pressure SMC to mold truck front ends is outlined. The production of cast steel molds is described.

481. Ampthor, F. J., Haoriliac, S., and Skogland, G. S., "Laboratory Factors for evaluating the Paintability of Press-Molded Low Shrink/Low Profile Polyester Parts," Ibid., Sec. 22-B.

Abs. The factors that influence the performance of painted parts when soaked in water at 90° for 10 days were studied. The influence of molding pressure, time, and temperature on paintability were investigated. Painting parameters, such as film thickness, and baking temperatures were also investigated. Properly molded and painted parts survived the test conditions.

482. Madine, T. V., "Problem Solving on Automotive SMC Panels through Quality Analysis," Ibid., Sec. 24-D.

Abs. Mechanical and aesthetic defects in SMC molded parts are described. A hasty description of a quality control program on SMC molded parts is presented.

483. Petersen, P., "New Procedures to Study the Flow Behavior of SMC," Ibid., Sec. 24-E.

Abs. NDT and destructive X-ray methods are described for analyzing the flow of SMC in molded parts. Contact print method and chemical silver coat plating are described for identifying defects.

484. Bastone, A. L., "Structural SMC-New Design Horizon for RP," *Plas. World*, Feb. 1978, pp. 64–67.

Abs. A good review of the OCF approach to structural SMC.

485. "FRP Wheels—How Soon Will They Roll?" *Plas. World*, Mar. 1978, pp. 52–54.

486. Miller, B., "Class 'A' Finish on SMC—How It Can Be Molded-in," *Plas. World*, Mar. 1978.

Abs. The GTR-GMMD development of in-mold coating is described.

487. Vacarri, J. A., "Urethane Coatings: Tough, Wear Resistant, Energy Savers," *Production Engineering* (Apr. 1978), pp. 40–42.

Abs. The six major types of urethane coatings are described. The coating vendors and their trade names are listed as well as a good discussion of the applications for automotive topcoats, marine coatings, and some special applications.

488. Ampthor, F. J., "Sink Reduction in SMC," *Plas. World*, May 1978, pp. 48–51.

Abs. Sink marks can be reduced by adjusting formulations, processing, and physical variables.

489. Sanders, B. A., "Engineering Properties of Automotive Fiber RP," Proceedings Soc. of Mfg. Engrs., Los Angeles, CA, Jun. 6–8, 1978, Paper EMR-78-04.

   Abs. This paper discusses the use of structural grade SMC in automotive parts as a weight-saving device. The properties of several different glass content compounds are presented.

490. Blatt, R. W., "Automated Processing for Compression Molding," Proceedings 4th SME Structural Conf., Los Angeles, CA, June 6–8, 1978, Paper EM-78-404.

   Abs. This paper describes SMC sheetmaking, compression press controls, molding heat management, loading mechanisms, material reactivity, and multiple station transfer lines.

491. Anon., "For Better Damage-Resistant Toughened SMC," *Plas. World*, June 1978, pp. 62–64.

   Abs. A rundown of the rubber-type polymer additives that are available.

492. Naitove, M. H., "Injection Molding Revitalizes Large FRP Auto-Parts Production," *Plastics Technology*, June 1978, pp. 79–84.

493. Anon., "Polyester Initiators: R. & D. Active in SMC-BMC," *Plas. Tech.*, Jul. 1978, pp. 124–128.

   Abs. A review of the catalysts for use in BMC-SMC is presented.

494. "Evaluating Glass Fiber Reinforcements," *Plas. Compounding*, Jul.–Aug. 1978, pp. 14–25.

   Abs. This article is a good review of the available glass fiber products and a short discussion of their applications.

495. Gallagher, R. B., and Kamath, V., "Organic Peroxide Review-Part 2: High Temperature Cure of Polyesters with Organic Peroxide and Azo Initiators," *Plas. Des. & Processing*, Aug. 1978, pp. 62–67.

   Abs. This is a good review of the various press molding catalysts that are available.

496. Cowley, M., "Spinning Is Tops," *Finishing Industries*, Sept. 1978.

   Abs. Electrostatic atomizers rotating at very high speeds are capable of dispensing high solids coatings and waterborne coatings that are difficult to apply by any other method.

497. Lambert, G. H., "Problems of Moulding SMC for the Automotive Paint Industry," Proceedings BPF RP Congress, Hotel Metropole, Brighton, England, Nov. 14–16, 1978, Paper 29, pp. 203–205.

Abs. This paper outlines the practical problems that have occurred in attempting to make an automotive component in which mechanical integrity and surface appearance are both critical.

1979

498. Auerbach, M., and Evans, D., "Repair of SMC," preprints SPI RP/CI 34th Ann. Mtg., New Orleans, LA, Jan. 29–Feb. 2, 1979, Ses. 1-E.

Abs. A two-component filled polyester repair system for voids, etc., and an air dry glaze for pin holes or porosity are described.

499. Magrans, J. J., and Ferrarini, J., "Unique Electrical and Mechanical Properties of ITP™," Ibid., Ses. 2-E.

Abs. This paper discusses the development of SMC systems using ICI's ITP™ resin system. Significantly improved electrical properties are claimed.

500. Ferrarini, J., Magrans, J., and Reitz, J., "New Resins for High Strength SMC," Ibid., Sec. 2-G.

Abs. This paper discusses the use of ICI's ITP™ resins for structural SMCs. Significant process advantages are claimed as well as improved flow in the mold and increased impact resistance.

501. Jacobs, E. F., and O'Hearn, T. P., "A New Resin and New Concept for High Performance Vinyl Ester Premix Molding," Ibid., Ses. 6-E.

Abs. A new vinyl ester resin that is thickenable by an organic thickening agent is described. Increased physical properties and added corrosion resistance are claimed.

502. Burns, R., and Pennington, D., "Study of Glass Fibre Degradation and Its Control During the Mixing of DMCs," Ibid., Ses. 7-D.

Abs. Fiber degradation occurs during the mixing of DMCs. Quick methods of accessing fiber degradation have enabled the rate of degradation to be equated to compound, glass fiber, and mixer variables. Hard glass types are less susceptible than are soft types.

503. Denton, D. L., "Mechanical Properties Characterization of an SMC-R50 Composite," Ibid., Sec. 11-F.

Abs. The mechanical properties of an R-50 composite over a temperature range of $-40°C$ to $121°C$ are reported.

504. Rapp, R. S., "SMC: The Effects of Isophthalic Polyester Processing," Ibid., Sec. 16-F.

Abs. The effects of 1-stage vs. 2-stage cooks, pressure processing, molecular weight variations, and the use of an esterification catalyst in processing isophthalic resins are discussed.

505. Burch, D. P., Ray, D. J., and Woodrich, R. H., "Design, Development, and Production of a One-Piece SMC Hood," Ibid., Sec. 18-A.

Abs. Abstract only in the preprints.

506. Nelb, R. G., "Auto Parts Manufacture—A New Dimension," Ibid., Sec. 18-C.

Abs. This paper describes the injection-molded BMC process installed at Bailey Div., USM Corp. for making large automotive parts.

507. Suter, L., "The First Production SMC Door," Ibid., Ses. 18-D.

Abs. The prototyping of an SMC door for Mack Truck is described.

508. Wood, E. L., Rockwood, L., Foster, J. E., and Nelson, G., "Increased Automotive RP Production and Quality through Automation of the SMC Process," Ibid., Ses. 18-E.

Abs. This paper describes the new Ford SMC plant in Milan. No photographs included.

509. Stanley, A., "Wheel Wells of Structural SMC-For Transportation Industry," Ibid., Ses. 21-C.

Abs. The use of continuous rovings in combination with chopped strands on an SMC machine are described.

510. Izzard, K. J., and Newton, G. P., "The Roll of Mixed Organic Peroxides in the Curing of DMC," Ibid., Ses. 24–B.

    Abs. The disadvantages of TBPB and TBPEH individually can be overcome by using mixtures of these two organic peroxides to cure DMC.

511. McCabe, M. V., "The Effects of Various Chemical Environments on the Flexural Properties of Molded SMC," Ibid., Ses. 22-C.

    Abs. Varying the sample size had very little effect on flexural property retention. Deionized water and alkali environments affected strength more than an acidic environment.

512. McLaughlin, P. V., McAssey, E. V., and Dietrich, R. C., "Non-Destructive Examination of Fiber Composite Structures by Thermal Field Techniques," Ibid., Ses. 22-D.

    Abs. This paper describes the use of thermography as a technique for examining large surface areas rapidly, without the need for data reduction and interpretation. Delaminations, blind surface impact damage, and surface cracks can be detected by infra-red thermal field techniques.

513. Gracrwski, A. S., Mandell, J. F., and McGarry, F. J., "Factors Affecting the Impact Resistance of SMC Materials," Ibid., Ses. 23-D.

    Abs. The effect of Paraplex™ P-13, rubber modifiers, cloth reinforcement, and Mylar™ fiber on the impact resistance of SMC molded panels is presented.

514. Schrantz, J., "High-Volume Urethane Applications," *Industrial Finishing* (Jan. 1979), pp. 18–21.

    Abs. The Chevrolet Flint, Michigan, system for automatically spraying a 2-mil coating of urethane primer on Monte Carlo and Malibu front ends is described. The low temperature cured primer (30 min. @ 180°F) provides a 2H-4H pencil hardness and effectively prevents solvent popping during assembly plant lacquer topcoat applications.

515. Anon., "Reinforced Plastics: What the Designer Is Looking For," *Plas. Design Forum*, Jan.–Feb. 1979, pp. 21–24.

Abs. The results of a survey of PDF readers on their RP requirements is presented.

516. Yee, G. Y., "Innovative SMC Process," SAE paper 790170, Cobo Hall, Detroit, MI, Feb. 26–Mar. 2, 1979.

Abs. The GM programmable force velocity control for compression presses, in-mold coating, an SMC line, mold heat transfer analysis, and a multi-station rotary compression molding line are discussed.

517. Nelb, R. B., "A New Process for FRP Auto Parts," Ibid., SAE paper 790209.

Abs. The injection molding of BMC automotive parts is described. Lower part weight, absence of sink marks, excellent part-to-part uniformity, and freedom from blistering at an elevated temperature are claimed.

518. "Plastics in Transportation: Slower Car Sales Won't Hurt Growth," *Plas. Tech.*, Feb. 1979.

Abs. Statistics on the use of FRP in trucks, buses, and cars are briefly described.

519. Yonagi, S., Hirai, Y., and Sakaguchi, S., "Curing Agents for SMC and BMC for Higher Productivity," *Kyoka Purasuchikusu*, Mar. 1979, pp. 96–102 (in Japanese).

520. Wilkinson, W. L., "GRP Processing: A Review of Progress and Future Developments," *Plas. Rub. Proc.*, Mar. 1979, pp. 1–9.

521. van Gasse, R. L., "Applications of Highly Filled DMC for the Building Industry in Europe," *Eur. Plas.*, Apr. 1979, pp. 41–43.

Abs. The use of DMC in preference to other polyester systems for building applications include roofing, cladding, and concrete building plans. The higher filler loading improves fire resistance.

522. Forger, G., "Current Trends in Electricals: the Case For FRP," *Plas. World*, Apr. 1979, pp. 69–71.

Abs. BMC electrical applications involving disconnect switches, breaker cases, and buss insulation are discussed.

523. "Plastics In Cars," *European Plas. News*, Apr. 1979, pp. 12–29.

Abs. A comprehensive review of the use of plastics in cars, with the emphasis on European cars, is given. Statistics are included.

524. Fock, K., and Kittl, R., "Practical Testing Methods for UP Sheet and BMC," *Materialpruefung,* May 1979, pp. 161–166 (in German).
525. Atkins, K. E., and Gandy, R. C., "Low Shrinking Thermosetting Molding Compositions Having Reduced Initial Viscosity," U.S. Patent 4,172,059, Oct. 23, 1979.
526. Kasischke, R. S., "The Accent Is on Injection Molding of BMC," Proceedings SPE NATEC, Nov. 6–8, 1979, Detroit, MI.
527. "Optimum Calcium Carbonate Selection for Reduced Density in SMC/BMC," Bulletin T-5-5a, Thompson-Weinman & Co., Cartersville, GA, 1979.

    Abs. Six formulations using Rohm & Haas Paraplex™ P-19D resin with combinations of calcium carbonate and glass bubbles are presented. Six BMC formulations using Marco GR-13024 resin, calcium carbonate, glass bubbles, and glass fibers are given together with physical properties.

1980

528. Ritter, J. R., "The Glass Phoenix," Preprints SPI RP/CI 35th Ann. Mtg., New Orleans, LA, Feb. 4–8, 1980, Ses. 1-A.

    Abs. This paper discusses the use of glass spheres as an additive in BMC and SMC to reduce viscosity, improve surface finish, improve toughness, etc.
529. Horner, A. H., Zaske, O., and Husain, K., "Injection Molding BMC-Develop a Manufacturing Window," Ibid., Ses. 2-A.

    Abs. A designed experiment was performed varying the molding temperature, molding pressure, and the compound velocity in a flat mold. The effect of fiber orientation was pronounced. It was found that the different molding temperatures did not affect physical properties; increased pressure gave higher physicals, as did shorter cure times. The injection velocity had no effect.

530. Leichtle, I. J., and Hinchman, G. H., "Continuous and Automatic Control of the SMC Process," Ibid., Ses. 2-B.

Abs. This paper describes the automated SMC process employed at Rockwell International.

531. Jackson, R. J., Farris, R. D., and Britigan, W. V., "A Thickenable Premium Vinyl Ester for SMC Applications," Ibid., Ses. 2-C.

Abs. This paper discusses a vinyl ester resin, Epocryl™, for use in structural SMC applications.

532. Cranston, J. J., and Reitz III, J. A., "SMC Molding Techniques for Optimized Mechanical Properties in Structural Applications," Ibid., Ses. 2-D.

Abs. This paper discusses the effect of mold charge placement in a flat sheet mold and of a prototype mold construction used to study the molding parameters of structural SMC.

533. Silva-Nieto, R. J., Fisher, B. C., and Birley, A. W., "Rheological Characterization of Polyester Resin SMC," Ibid., Ses. 2-E.

Abs. This paper presents a theoretical and experimental analysis of the rheological behavior of SMC.

534. Finsel, J. R., and Younge, K. R., "Hybrid BMCs-FRP Thermosets Displace Traditional Molding Materials," Ibid., Ses. 2-F.

Abs. The markets and applications for BMC continue to grow. Injection molded BMC further reduces costs and improves performance.

535. Golemba, F., "A New Modified Rubber Additive to Improve Toughness of Polyesters," Ibid., Ses. 5-A.

Abs. This paper discusses a modified rubber ester, which when added to polyester resins improves fracture touchness while maintaining or improving mechanical and physical properties.

536. Gruskiewicz, M., and Collister, J., "Analysis of the Thickening Reaction of an SMC Resin through the Use of Dynamic Testing," Ibid., Ses. 7-E.

Abs. This paper discusses the change in viscoelastic

response of a polyester resin during the course of thickening with MgO. Rheological measurements were made to describe the effects of thickening as functions of time and MgO concentration.

537. Grant, I., Feltzin, J., and Flint, A. J., "Dynamic Mechanical Properties of ITP™ Resin Based SMC Important in Automotive Design," Ibid., Ses. 8-B.

Abs. The advantages of the ICI America's novel polyurethane thickening process for making SMC are reviewed. Static and dynamic mechanical performance data are developed for polyurethane thickening and are compared with conventional thickening systems.

538. Ehnert, G., "Further Prospect of Market Requirements and SMC Developments for Automotive Applications in Europe," Ibid., Ses. 8-D.

Abs. This paper reviews the recent development activity in SMC in Europe aimed at the automotive industry. Some of the problems that have prevented faster use of FRP are discussed.

539. Greenzweig, J., Moulson, T., and Quick, W., "Molecular Weight Control in SMC Polyester Resins," Ibid., Ses. 10-D.

Abs. This paper points out that molecular weight distribution control is essential in producing batch-to-batch consistency in SMC resins. The best way to ensure this consistency is by gel permeation chromatography as a quality control tool.

540. Dickson, R. T., "HSMC Radiator Support," Ibid., Ses. 13-A.

Abs. This paper investigates the feasibility of replacing the typical multi-piece steel weldment radiator support, with a less complex compression molded HSMC part. A three-piece adhesively bonded assembly has been developed that matches the performance of a steel part and offers a 7.5 pound weight advantage at a slight cost penalty.

541. Englehart, W. H., "The Dependency of Part Performance on Material and Process Controls," Ibid., Ses. 7-A.

Abs. This paper deals with the relationship of part performance to SMC materials and process control. It shows how the controls were established on an SMC truck hood to insure a high degree of reliable performance.

542. Allen, R. C., "Low Density SMC: Evolution through Challenge," Ibid., Ses. 7-B.

    Abs. This paper discusses two alternatives of weight reduction; down gauging of part thickness and lower density parts. Down gauging appears best suited for appearance parts where impact strength is not a necessity. Some cost penalty is involved in using lower density SMC.

543. Bollen, P. S., Degrassi, A., and Sacks, W., "Nylon Support Films for the Manufacture of SMC Products," Ibid., Ses. 7-C.

    Abs. Films of modified Nylon 6 resin at 1 mil thickness are 3,500 times less permeable to styrene monomer compared with 2 mil low-density polyethylene.

544. Tallback, D. N., "TMC™—Uses and Applications," Ibid., Ses. 7-D.

    Abs. This paper discusses the advantages and uses of thick molding compound.

545. Auerbach, M., "Repair of SMC-Phase II," Ibid., Ses. 16-B.

    Abs. This paper describes materials and techniques for repairing some SMC defects.

546. Foreman, P., Dunn, B., and Eagle, G., "New Urethane Adhesives and Adhesive Sealants for Bonding SMC," Ibid., Ses. 16-C.

    Abs. A complete system for the structural bonding and final sealing of plastic and metal assemblies is described. Two-part urethanes are used, which cure at room temperature.

547. Mackey, B. A., "How to Drill Precision Holes in Reinforced Plastics in a Hurry," *Plastics Engineering* (Feb. 1980) p. 22.

    Abs. Existing tungsten carbide tooling uses the right material but the wrong design for cutting reinforced plastics. The tool can tolerate the abrasion, but the plastic can not withstand the heat the tooling is designed to create for cutting cast iron. New tungsten carbide tooling designed with positive rake not only reduces unwanted heat but also cuts through the fibers more cleanly and up to four times faster.

548. Wood, A. S., "Now BMC Is Easier to Handle and Tougher," *Mod. Plas.*, Mar. 1980, pp. 64–67.

    Abs. Continuous mixing/injection molding systems, teamed up with materials improvements, open new opportunities for economical mass production of high-performance components, particularly appearance parts.

549. "Polyceramics: A New Low-Styrene RP," *Mod. Plas.*, Jul. 1980, p. 14.

    Abs. News release by Tanner Chemical Company concerning a hybrid of polyester resin and ceramics.

550. Cope, D. E., "Hydrophobic Filler Wetting Yields Better Composites," *Mod. Plas.*, Jul. 1980, pp. 60–66.

    Abs. This article discusses the effects of hydrophobic wetting agents and describes the filler surface modification techniques.

1981

551. Luzier, W. D., and Howatineck, F. J., "Improved SMC Systems for Electrical Applications," Preprints SPI RP/CI 36th Ann. Mtg., Washington, D.C., Feb. 16–20, 1981, Ses. 1-A.

    Abs. An SMC system using ITP™-1054 vinyl ester resin showed superior electrical and water absorption properties compared with an MgO thickened system.

552. Hurst, A. T., Smith, S. A., and Reitz, J. A., "Enhanced Toughness through Urethane Thickened SMC," Ibid., Ses. 1-B.

    Abs. This paper explores the effects of SMC formulation variables on toughness. A statistically designed experiment compares an MgO and a urethane-thickened system.

553. Howes, W. C., and Smiley, L. M., "Maximizing the Performance of Hollow Microspheres in Low Density SMC," Ibid., Ses. 1-C.

Abs. This paper considers mechanical properties, cost, and density of LD-SMC. Microspheres are used to lower the material density.

554. Voeks, S. L., and Kallaur, M., "Tough SMC Resins," Ibid., Ses. 1-D.

Abs. Several polyester resins are evaluated for toughness by comparing standard impact test values and the GM matrix strain method. A new resin, Stypol™ 40-3919, shows up well on the comparison.

555. Horner, A. H., Zaske, O., and Rapp, R., "Injection Molded BMC-Influence of Process Limits on Product Physicals," Ibid., Ses. 5-E.

Abs. Only an abstract appears in the preprints. A hard copy of the paper was obtained from the authors.

556. Greening, T., "International Harvester SMC Facility Design," Ibid., Ses. 6-A.

557. Foster Jr., J. E., "SMC Paste Metering," Ibid., Ses. 6-D.

558. Das, B., Loveless, H. S., and Morris, S. J., "Effects of Structural Resins and Chopped Fiber Lengths on the Mechanical Properties and Surface Properties of SMC Composites," Ibid., Ses. 10-B.

560. Myers, F. A., and Nyo, H., "Structure-Performance Relationship for SMC," Ibid., Ses. 11-A.

Abs. The results of a scanning electron microscopic study of SMC are presented.

561. Liebold, R. W., "High Strength Bumpers for Cars in Series Production with SMC-C," Ibid., Ses. 11-B.

Abs. The development of the HSMC bumper on the Peugeot 505 is described.

562. Seamark, M. J., "Facelifting the World's First All SMC Clad Truck Cab after Five Years Production—A Unique Case Study," Ibid., Ses. 11-C.

Abs. The history of the first European all SMC truck cab is presented together with the details of the panel design changes after five years' production.

563. Denton, D. L., "Effects of Processing Variables on the

Mechanical Properties of Structural SMC-R Composites," Ibid., Ses. 16-A.

Abs. Designed experiments were conducted to check the effect of process variables (SMC machine speed, chopper speed, mold coverage, mold temperature, pressure, etc.) on the mechanical properties of SMC-R flat sheets.

564. Panter, M. R., "The Effect of Processing Variables on Curing Time and Thermal Degradation of Compression Molded SMC," Ibid., Ses. 16-F.

Abs. A study of the effect of mold temperature, preheating, and mold material on the cure time and thermal degradation of SMC plates and cylinders is presented.

565. Greenweig, J. E., and Nelson, J. L., "Low Profile Additive for Flame Retardant SMC," Ibid., Ses. 18-A.

Abs. A low-viscosity LP additive permits high ATH loadings to BMC and SMC.

566. Pinkowski, N. J., "New High Density-Low Viscosity Zinc Stearate for Use in Polyester Molding Compounds," Ibid., Ses. 18-B.

Abs. A new dense zinc stearate mold release agent is described. Lower initial soup viscosities, higher filler loadings, and lower dusting are claimed.

567. Fintelmann, C., Zaske, O., and Horner, A. H., "A Cost Effective LP Additive for Automotive and Industrial Applications," Ibid., Ses. 18-C.

Abs. S-846, a saturated polyester solution in styrene is described and is compared in several formulations with other LP additives. A good discussion of shrink control is provided. S-846 is claimed to provide process, product, and cost advantages in BMC and SMC.

568. Trudeau, E. G., and Lindsay, M. W., "SMC-the Second Challenge," Ibid., Ses. 21-B.

Abs. A discussion of the Ford SMC/HSMC light truck tailgate is presented.

569. Bettac, W. E., and Gray, Jr., E., "Squeeze Molding Technique as a Means of Simulating SMC Molded Part Performance," Ibid., Ses. 21-D.

Abs. An SMC prototype automotive hood made on a hand lay-up fixture is described.

570. Bence, I. J., Bloom, R. C., and Gray, Jr., E., "Design, Finite Molding, and Structural Performance of a One-Piece SMC Auto Hood," Ibid., Ses. 21-E.

Abs. This paper presents the design details and structural analysis approach used to develop a 1-piece SMC automotive hood.

571. Gerard, J., Keown, J., and Loyal, B., "Designing and Prototyping A European Automotive SMC Door," Ibid., Ses. 21-G.

Abs. The design, prototyping, and testing of an SMC door for a Peugeot 305 automobile is presented. Lower weight and fewer pieces were realized in the SMC door as compared with steel.

572. Zaske, O., Fintemann, C., and Wuh, J., "The Effect of ATH Characteristics on the Thickening Behavior of ATH-Filled SMC Pastes," Ibid., Sec. 8-F.

Abs. The SPI thickening test reproducibility is discussed. By increasing sample size, adding the thickener as a dispersion, and using a higher shear mixer reproducibility appears to be improved.

573. Sherer, D. L., and Ashley, M., "Acoustic Emission Testing of HMC™ Composite," Ibid., Ses. 23-B.

Abs. This paper demonstrates that acoustic emission testing may be used as a screening method for HMC parts in the flexural mode.

574. Menges, G., and Derek, H., "Flow Testing of SMC and Its Practical Application," Ibid., Ses. 23-C.

Abs. An instrumented mold is described in this study. A basis for designing charge patterns is discussed based on flow behavior in a hot compression mold.

575. Lee, B. L., and Howard, F. H., "Effect of Matrix Toughening on the Strength and Strain Properties of SMC," Ibid., Ses. 23-E.

Abs. The resistance of SMC to matrix cracking was found to be governed by crack toughness of the matrix resin. Crack toughness of the resin can be improved by including rubber particles in the polyester resin.

576. Lee, L. J., Marker, L. F., and Griffith, R. M., "The Rheology and Mold Flow of Polyester SMC," *Polymer Comp.*, Oct. 4, 1981, pp. 209–218.

Abs. The flow properties of SMC are measured by several rheometers. Shear viscosities of SMC paste can be fitted by the Carreau viscosity equation. The deformation of glass-filled compounds under shear forces was quite different from those under extensional forces. When the shear forces were applied, the compound deformed like a deck of cards being slid. The flow of SMC compound during molding showed that the surface layers flowed farther than the inner layers when the mold surface was hot.

577. Scharfenberger, J. A., "The Use of Higher Speed Electrostatic Disk and Bell systems for the Application of high Solids and Waterbourne Coatings," Technical Report TL-81-03, Ransburg Electrostatic Equipment, Indianapolis, IN.

    Abs. With the development of waterbourne and high solid coatings to meet environmental legislation new application equipment was necessary. Rotating electrostatic atomizers often referred to as "disks" or "bells" with variable speed drives operating at very high speeds can be satisfactorily used to dispense such coatings.

1982

578. Thompson, E. J., Alberino, L. M., and Farrissey, W. J., "Isocyanate Thickened SMC," Preprints SPI RP/CI 37th Ann. Meeting, Washington, D.C., Jan. 11–15, 1982, Ses. 1-A.

    Abs. This paper presents a discussion of isocyanate-thickened SMC and compares it with conventional MgO thickened systems.

579. Horner, A. H., Brill, R., and Zaske, O., "Factors Influencing SMC Paste Viscosity and Control—Some Second Thoughts," Ibid., Ses. 1-B.

    Abs. The results of a full factorial experiment involving 3 levels of temperature, 3 levels of water, 2 levels of thickener, and time is reported with the paste viscosity being the dependent variable.

580. Myers, F. A., "Impact Response of SMC/BMC Composites," Ibid., Ses. 1-C.

Abs. The impact response of SMC and BMC materials is examined by employing instrumented drop weight tests on flat samples. The variation in load and energy with tip geometry, plate thickness, and impact velocity is discussed.

581. Silverman, E. M., "Material Properties of Chopped Glass-Fiber/PE SMC," Ibid., Ses. 1-D.

Abs. The influences of material heterogeneity and anisotropical fiber spacing, local fiber content, fiber orientation, and fiber/matrix adhesion on the mechanical properties of HMC are examined.

582. Emrich, P., and Riddell, R. M., "Effects of Preworking Glass Rovings on Matched Metal Molded RP," Ibid., Ses. 1-E.

Abs. This paper discusses a device for applying tension to rovings for use on preform machines and possibly for use on SMC machines as well. A glass evaluation is also included.

583. Boot, R. J., Meyer, R. W., and Copeland, J. R., "Replacement of Glass Fiber by Fiber Grade Wollastonite in a Polyester BMC Compound," Ibid., Ses. 6-C.

Abs. This paper discusses replacing short glass fibers in a BMC with fiber grade wollastonite.

584. Carroll, W., and Fackler, M., "Low Modulus SMC for Automotive Applications," Ibid., Ses. 7-A.

Abs. This paper discusses low modulus, damage-resistant SMC automotive parts that retain the traditional properties of high strength, light weight, low profile, corrosion resistance, weatherability, paintability, and high impact strength.

585. Atkins, K. E., Gandy, R. C., and Gentry, R. R., "SMC-Thinner-Smoother-Tougher; A New Generation of LP Additive," Ibid., Ses. 7-C.

Abs. The use of LP XLP-4514 and XLP-8521 to produce thinner, smoother, and tougher SMCs for automotive body panels is discussed.

586. Gardner, I. L., Gursky, L., and Fasco, J. V., "Conjugated Diene Butyl—A Unique LP Toughening SMC Additive," Ibid., Ses. 7-D.

Abs. Properly compounded and mixed SMC containing CDB can achieve class A surfaces compared with conventional LP additives while maintaining excellent physical properties.

587. Hurst, A. T., "Measurement Aspects and Improvement of Surface Profiles in Thin Gauge Molded SMC," Ibid., Ses. 7-E.

Abs. This paper describes a new method of measuring surface profile.

588. Quick, J. R., and Mate, Z., "Conductive Nonwoven Fiber Mats for Electromagnetic Interface Shielding in SMC Parts," Ibid., Ses. 7-F.

Abs. Nonwoven fiber mats prepared from conductive fibers can be used to provide electromagnetic interference shielding in SMC molded parts.

589. Ross, G., "Advances in SMC Process Control," Ibid., Ses. 13-A.

Abs. This paper reviews concepts being investigated to reduce the variability of the SMC process.

590. Brueggemann, W. H., "Powder Premold Coating for SMC/BMC," Ibid., Ses. 13-B.

Abs. This paper discusses a relatively new process of powder premold coating a hot mold before charging the molding compound to the mold. Data are reported on extensive laboratory and plant trials.

591. Zaske, O. C., Fintlemann, C., and Horner, A. H., "Plant Processable Rubber Modified SMC," Ibid., Ses. 13-C.

Abs. S-846, a saturated polyester resin, can moderate the rubber introduced viscosity increases in SMC pastes according to this paper. The pastes then are readily processed on standard equipment using standard procedures.

592. Cadman, R. K., Nelson, M. F., and Chen, F. H., "Structural Analysis of a Fiber-Reinforced Plastic Truck Cab," Ibid., Ses. 16-B.

Abs. This paper describes the structural analysis phase

of a project at GM to design, analyze, build, and test an experimental FRP cab for the GMC General truck.

593. McIver, G. M., and Thompson, D., "Structural Adhesive Determination for the GMC Plastics Cab," Ibid., Ses. 16-C.

Abs. This paper describes the evaluation of 3 adhesive systems on 4 structural laminates for a totally bonded plastic truck cab.

594. Fulmer, R. W., "Advanced Structural and Hybrid SMC," Ibid., Ses. 20-C.

Abs. This paper summarizes the development work to date on ultrahigh performance structural SMC and to point out some of the possible applications of this material.

595. Fletcher, C. W., "An Examination of the Mechanism by Which Stearates Act As Internal Mold Release Agents in Matched Metal Die RP Molding," Ibid., Ses. 22-F.

Abs. The theory of how stearates act as mold release agents is examined through the use of scanning electron microscopes using X-ray energy dispersion analysis to identify the degree of dispersion of the stearate throughout the matrix as opposed to the concentration on the part surface.

596. Kirsch, P. A., and Champion, R. L., "Development Effort of SMC Parabolic Trough Ends," Ibid., Ses. 26-C.

Abs. This paper describes the development effort to produce SMC parabolic trough solar reflector panels.

597. Burns, R., *Polyester Molding Compounds*, Marcel Dekker, Inc., New York, 1982, 335 pp.

Abs. This is one of the few books exclusively devoted to polyester molding compounds. Chapter 1 deals with definitions, make-up of the market, compound usage by years and markets. Chapter 2 is an extensive coverage of the raw materials involved. It constitutes approximately a third of the book. Chapter 3 deals with the preparation of molding compounds. British materials and equipment are emphasized. Chapter 4 is an extensive discussion of the rheology of molding compounds. Chapter 5 covers the physical properties of molded parts, the effect of different grade glass strands, the effect of fiber length, RFI shielding, etc. Chapter 6 discusses newer compound developments, such as low-pressure compounds, TMC™, and high-strength compounds. Chapter 7 deals with compression and injection molding of SMC and BMC. Chapter 8 covers finishing and secondary operations. Chapter 9 concludes with a case study of the ERF FRP truck cab.

1983

598. McClusky, J. J., and Jutte, R. B., "Development of SMC for Automotive Body Panel Use," Preprints SPI 38th RP/CI Ann. Mtg., Houston, TX, Feb. 7–11, 1983, Ses. 1-A.
    Abs. An OCF study has shown that a high-quality surface finish SMC can be produced. A 90-second cure time was demonstrated on a 500-part run.

599. Ziehm, J. J., "One-Step Priming System for SMC," Ibid., Ses. 1-D.
    Abs. The Sherwin Williams in-mold coating system is discussed.

600. Simko, T. D. and Pladek, O., "Fillers as Reinforcement Replacements in Various SMC Formulas," Ibid., Ses. 3-A.

601. Arbit, H. A., and Howes, W. C., "Solid Glass Spheres as a Functional Additive for BMC," Ibid., Ses. 3-C.

602. Lede, B. L., Howard, F. H., and Rowe, E. H., "Effect of Matrix Toughening on the Crack Resistance of SMC Under Static Loading," Ibid., Ses. 9-A.
    Abs. This paper discusses the rubber-toughening of SMC.

603. Kallaur, M., and Matous, J. J., "A Novel Thermosetting Acrylic Resin System for Improved SMC Applications," Ibid., Ses. 9-G.
    Abs. An interpenetrating network system is described that provides improved resistance to water, temperature, and fatigue.

604. Taylor, E. J., "A Mixture Experiment Approach to the Formulation of an SMC Low-Profile Resin System," Ibid., Ses. 9-C.

Abs. An SMC low profile resin system was subjected to a mixture experimental design that provided improved physical and mechanical properties and improved surface properties.

605. Horner, A. H., Brill, R. N., Fintelmann, C., and Zaske, O. C., "Consider SMC Paste as a System—The Key to Success in SMC Paste Process Control," Ibid., Ses. 9-D.

Abs. This paper points out that the factors that influence thickening are time, temperature, water content, thickener content, and, above all, the interrelationships that develop among these various factors.

606. Himebaugh, D. C., and Newman, S., "Analytical Techniques for Measuring the Void Content of SMC Sheet," Ibid., Ses. 9-E.

Abs. This paper discusses 3 test methods used to access void content of SMC paste and sheet. The apparatus used are described. SMC paste was found to contain 1.1-2.4 cm^3 air per 100 grams. By degassing, that value was reduced to essentially 0.

607. McCarghy, N., and Collister, J., "Introduction of Novel Pelletized Thermoset Polyester Molding Compounds," Ibid., Ses. 9-F.

Abs. This paper describes two grades of Premi-Dri™ pelletized molding compounds, together with physical properties, and molding notes.

608. Doyle, M. J., and Gardner, I. J. "Toughening of Polyester Resin SMC with Conjugated Diene Butyl Elastomer," Ibid., Ses. 19-G.

Abs.

609. Owens, J. H., and Johnson, H. C., "Automated Systems for Application of Premold Coatings," Ibid., Ses. 20-F.

Abs. Powdered premold coating equipment is described, and the properties of this system are compared with conventional systems. The system is explained in simple sketches, and its reported cost savings are explored.

610. Peters, J. B., "Present and New Technology of Coating and Finishing of RP," Ibid., Ses. 23-D.

Abs. Conventional and alternative methods of finishing FRP products are discussed.

611. Lasell, D. M., and Lowry, J. R., "Mechanical Property Improvement of Injection Molded Premix," Ibid., Ses. 24-E.

Abs. This paper reports on efforts at the Bailey Corp. to reduce the mechanical property loss of injection molded BMC caused by reinforcement degradation.

612. Beatty, R., "Paint Curing Process Will Have Immense Impact," *ert* (May 1983), pp. 10–12.

Abs. The Vapocure process in which a coating is cured at room temperature by exposing the coating to a catalyst in vapor form is described. Several applications are described as well as some pilot lines.

1984

613. Willis, C. L., Halper, W. M., and Handlin, Jr., D. L., "Elastomer Modified SMC," Preprints SPI RP/CI 39th Ann. Mtg., Houston, TX, Jan. 16–19, 1984, Sec. 2-C.

Abs. This paper reports on a morphological study of a new Shell elastomeric additive, GX-1855.

614. Horner, A. H., and Brill, R. N., "Develop Thickening Behavior Conditions to Insure SMC Process Control," Ibid., Ses. 8-A.

Abs. This paper discusses a full factorial experiment involving 3 levels of water and 4 levels of thickener to develop a set of boundary conditions for SMC paste.

615. Canning, J. L., "Formulating Polyester Molding Compounds through Use of the 'Resin Demand' Procedure," Ibid., Ses. 8-b.

Abs. A simple, easy-to-perform "resin demand" test procedure is described for use on a molding compound system.

616. Johnson, H. C., and Owens, J. H., "Commercial Developments in Premold Powder Coating of RP Substrates," Ibid., Ses. 8-D.

617. Speelman, D., and Wildman, G., "The Effect of Fillers on the Color of Plastic Composites," Ibid., Ses. 12-F.

Abs. This paper reports on the effect of colors on the most commonly used filler systems in SMC formulations. Similar brightness values for the same filler types give similar colors. Similar brightness values between filler types give wide variations of color. The lighter colors show wider differences than do darker colors.

618. Arakawa, K., Haraguchi, S., Iwami, E., and Nomaguchi, K., "Newly Developed BMC Technology for Electrical Parts," Ibid., Ses. 14-A.

Abs. This paper discusses improved processing in the manufacture of electrical power transformers that reduces costs, reduces styrene emissions, and uses lower temperatures and pressures in the molding operation.

619. Fekete, F., and Knowles, R. B., "Pelletized Polyester Molding Materials," Ibid., Ses. 14-C.

Abs. This paper describes the properties and end-use applications of a series of new generation free-flowing, solid, pelletized, reinforced, molding compounds.

620. Hoffman, J. D., Thompson, P. E., Miller, J. S., Chandalia, K. B., and Leung, W. D., "Novel SMC Formulation for Class A Finishes," Ibid., Ses. 14-D.

Abs. This paper describes the use of Uralloy™ LP 85-05 additive in SMC to produce materials with superior toughness and surface finish.

621. Sirabian, S., Eise, K., and Curry, J., "Multi-Screw Compounders for Reinforced Plastics Production," Ibid., Ses. 17-A.

Abs. This paper describes a multi-screw extruder for compounding pellatized molding compounds.

622. Gibson, A. G., and Williamson, G. A., "Polyester Premix Injection Rheology: Improved Properties through Reduced Process Work," Ibid., Ses. 17-B.

Abs. There is a loss of properties when BMC is injection molded. This paper describes improved designs for screws and nozzles that reduce the work input to the compound and improve the compound properties.

623. "Fiero Requires Unique Paint Line," *Industrial Finishing* (June 1984), pp. 16–20.

Abs. This article describes in detail the equipment used and painting procedures for the Pontiac Fiero automobile.

624. Schrantz, J., "Vapor Cure," *Industrial Finishing* (Feb. 1984), pp. 26–29.

Abs. This paper discusses a relatively new technology of using a vapor to rapidly cure coatings at room temperature. Some of the companies involved in this technology are mentioned as well as some typical applications.

625. Schrantz, J., "Painting the Corvette," *Industrial Finishing,* (Mar. 1984), pp. 18–22.

Abs. The basecoat-clearcoat finishing system used on current Corvette bodies is described.

1985

626. Atkins, K. E., and Gandy, R. C., "Polyester Modifiers: Further Advances in SMC Technology," SPI RP/CI 40th Ann. Mtg., Atlanta, GA, Jan. 28–Feb. 1, 1985, Session 1-E.

Abs. Neulon polyester modifier A is described as a higher-temperature-resistant material for automotive SMC, which provides molded parts with a greater than normal blister resistance. Improved surface formulations are presented, and the ability to accept in-mold coating of molded parts is claimed.

627. Lee, D. S., and Han, C. D., "Chemorheology of Low-Profile Polyester Resins," Ibid., Ses. 1-F.

Abs. The results of an experimental study to investigate the curing behavior of a LP resin system is discussed. Two different thermoplastic additives were used; namely, a PVAc and polystyrene-co-butadiene. The latter material has a shorter gel time than does the PVAc compound.

628. Miller, W. P., "A Dry Process Calcium Carbonate Filler for Thermoset Reinforced Polyester Compounds," Ibid., Ses. 3-E.

Abs. Dry processed calcium carbonate is compared with wet ground calcium carbonate, and initial non-thickened, thickened, and matured viscosities are essentially the same. The use of dry ground calcium carbonate is discussed in BMC, SMC, TMC, and LSW systems.

629. Joneja, S. K., and Newaz, G. M., "Mechanics and Test Methods of SMC Bond Adhesion," Ibid., Ses. 5-E.

Abs. This paper discusses a study of adhesively bonded SMC joints. It was found that a wedge test is an effective method of determining the performance of bonded joints.

630. Yen, S. C., and Morris, D. H., "Accelerated Characterization of the Creep Behavior of SMC-R50," Ibid., Ses. 5-F.

Abs. The creep response of SMC-R50 under different thermomechanical conditions is reported in this paper. The feasibility of predicting the long-term creep response using the Time Temperature Superposition Principle is discussed.

631. Kardentz, P. D., Messick, V. B., and Craigie, L. J., "Unique Vinyl Ester Provides Solutions to Tough Problems," Ibid., Ses. 6-C.

Abs. This paper discusses Derakane™ 8084 vinyl ester resin and explains how the new resin may expand application areas for thermoset resins.

632. Zaske, O. C., Wang, M., and Wuh, J., "The Effect of ATH Characteristics on the Thickening Behavior of ATH Filled SMC Pastes," Ibid., Ses. 8-F.

Abs. This paper reports on the results of a study of 7 ATH fillers. It was found that particle size distribution had the most effect on thickening. Chemical origin and differences in trace impurities had little effect on thickening.

633. Fekete, F., "Glass Reinforced, Pellatized, Granular and Nodular Type Unsaturated Polyester Thermoset Molding Compounds—A Status Review," Ibid., Ses. 9-A.

Abs. This paper covers a patent and market review of the various forms of glass-reinforced thermoset molding compounds that have been made in pellet, granular, modular, flake, and powder forms.

634. Wu, Y. T., "Super Wear Resistant BMC Reinforced with Aramid Fibers," Ibid., Ses. 9-B.

Abs. This paper discusses the use of Kevlar™ in BMC to provide equivalent mechanical and electrical properties and superior wear resistance as compared with glass-reinforced compounds.

635. Lynskey, B. M., and Robertson, F. C., "Modified SMC for Thick Section Components," Ibid., Ses. 9-D.

Abs. This paper discusses the use of inhibitors, retarders, secondary monomers, and alternative initiators to reduce the peak exotherm temperature of SMC without affecting the cured resin properties.

636. Liebold, R., "Injection Molding of Large-Area Glass Fiber Reinforced Components—A Comparison with Compression Molding from the Processor's Point of View," Ibid., Ses. 9-E.

Abs. This paper discusses the use of injection molding to mold large area parts and compares those parts with similar ones made by compression molding.

637. Fekete, F., and Martinez, V., "The Influence of Specific Processing Variables on the Physical Properties of SMC Molded under Restricted Flow Conditions," Ibid., Ses. 9-F.

Abs. This paper discusses the influence of specific processing variables on the physical properties of SMC molded under restricted flow conditions.

638. Sedlatschek, R. L., Wiehl, F. R., and Rutherford, J. L., "The Glass Reinforced Composite Sewing Machine Structure—Its Design and Development," Ibid., Ses. 13-B.

Abs. This paper describes the steps involved in designing and putting into production an injection molded thermoset polyester sewing machine.

639. Zilg, G. L., and Block, M., "Empirical Design Parameters for Self-Threading Fasteners in Injection Molded RP," Ibid., Ses. 13-D.

Abs. Thread-cutting screws are preferred over thread-forming screws, based on superior pull-out strength.

The findings for strip torque vs. length of thread engagement, strip torque vs. hole diameter, clamp force vs. applied torque, and breakaway torque vs. applied torque are reported.

640. Edwards, H. R., "Using Micro Analytical Techniques to Guide Development of a Class A Molding Compound," Ibid., Ses. 16-A.

Abs. This paper explores the use of acoustic emmissions to predict mechanical properties of molding compounds and the use of electron microscopes to observe the dispersion of elastomers that contribute to surface profile and the surface micropores upon which paint popping is blamed.

641. Schmelzer, E., Menges, G., and Cherek, H., "SMC-Processing Up the Learning Curve," Ibid., Ses. 16-B.

Abs. This paper discusses the use of a microcomputer to control the external sequence and peripheral equipment involved in molding SMC. Mold temperature has a significant influence on cure time.

642. Kanagendra, M., and Fisher, B. C., "Process Interactions for SMC Compression Molding Under Microcomputer Control," Ibid., Ses. 16-C.

Abs. This paper describes the design and use of a digital control system for a compression press and the data collected during the investigation of the interactions that occur between process variables and the rheology, heat history, and curing reaction within the mold.

643. Horner, A. H., and Brill, R. N., "Some Factors Influencing Polyester Resin Behavior during the SMC Thickening Reaction," Ibid., Ses. 16-D.

Abs. This paper considers the thickening process from the resin viewpoint. Using a known thickenable resin, the authors discuss its thickening behavior as a function of time, temperature, and thickener concentration. Indications are that the thickening reaction may be closely followed and predicted by monitoring the viscosity and acid number of the system as a function of time and temperature.

644. Costigan, P. J., Fisher, B. C., and Kanagendra, M., "The Rheology of SMC During Compression Molding, and Resultant Material Properties," Ibid., Ses. 16-E.

Abs. This paper reports the findings of an investigation of the effects of SMC molding variables. The type of flow in the mold is determined by process conditions. Tensile properties of molded plagues did not differ significantly for the different process conditions except when the charges were preheated.

Annotated Bibliography Author Index by Item Number

Ackley, R. H., 389
Ahrberg, W. R., 18, 24, 29, 34
Alberingo, L. M., 578
Allen, R. C., 542
Alspac, H. S., 133
Ampthor, F. J., 242, 300, 433, 472, 478, 481, 488
Anderson, T. E., 182
Arakawa, K., 618
Arbit, H. A., 601
Ashley, M., 573
Atkins, K. E., 270, 286, 372, 414, 424, 431, 525, 585, 626
Auerbach, M., 498, 545
Austin, T. E., 91, 281
Austin, V., 385
Backderf, R., 339
Bartkus, E. J., 155, 184
Bassford, F. C., 445
Bassler, R. B., 92, 100
Bastone, A. L., 484
Bauer, S. H., 306, 334, 346, 386
Baum, M., 174
Beatty, R., 612

Bence, I. J., 570
Berger, S. E., 414, 424
Bettac, W. E., 569
Biefield, L. P., 26
Bigelow, M. H., 1, 8
Birley, A. W., 533
Blanchard, B., 162
Blatt, R. W., 490
Block, M., 639
Bloom, R. C., 570
Blount, W. W., 246, 260, 338
Blue, E. B., 203
Boeker, B. E., 96
Bollen, P., 543
Boot, R. J., 583
Borro, E., 13
Bowyer, E. G., 449
Boyd, L., 370
Bradish, F. W., 240, 377, 447, 470
Bradley, I. G., 354
Bradley, R. H., 271
Brill, R. N., 579, 605, 614, 643
Britigan, W. V., 531
Brochne, F., 159

Brueggemann, W. H., 590
Burch, D. P., 505
Burer, E., 434
Burns, R., 406, 430, 457, 502, 597
Burton, G. W., 79
Busch, W., 234
Butler, J., 98, 453
Calderwood, R. H., 43, 76, 78
Calendine, R. H., 246, 338
Callahan, M. L., 348
Campana, T., 149
Canning, J. L., 181, 615
Carroll, W., 584
Caramante, D. E., 92, 100
Carter, N. A., 193, 202
Cawthorne, D., 319
Champion, R. L., 596
Chandalia, K. B., 620
Charters, C. A., 391
Cherek, H., 641
Chiesi, F., 455
Clavadetcher, D. J., 275, 447, 466
Coats, C. E., 438
Charters, C. A., 391
Colao, J. J., 40
Collister, J., 336, 439, 536, 607
Comey, J. F., 359
Comstock, L. R., 270, 286
Conley, D. O., 280
Cooper, T. I., 374
Cope, D. E., 550
Copeland, J. R., 583
Cordts, H., 167

Corneliussen, L., 398
Costigan, P. J., 645
Cowley, M., 496
Craigie, L. J., 631
Cranston, J. J., 532
Crenshaw, J. B., 33
Cross, J., 367
Culweir, E. F., 21
Currier, G. J., 259
Curry, J., 621
Cutler, N. A., 55, 75
Cutshall, J. E., 248
Czarnomski, J. J., 192
Das, B., 558
Dastur, S. M., 325
Davies, E. A., 187, 198
Davies, J. D., 46
Davies, J. H., 246
Davies, J. W., 388, 437
Davis, Sr., C. J., 235
Deis, W. H., 127, 171, 288, 323
Degrassi, A., 543
Deller, T. E., 350, 412
Denton, D. L., 503, 563
Derek, H., 574
DeVore, H. W., 2
Dickson, R. T., 540
Diebold, W. R., 92
Dietrich, R. C., 512
Dietz, A. G., 11
Dix, J. S., 475
Doob, H., 60
Doyle, M. J., 608
Doyne, R. F., 32
Dreger, D. R., 316
Drew, E. W., 268

DuBois, J. H., 25
Dunn, B., 546
Duprez, H. J., 48
Dzik, C. J., 250
Eagle, G., 546
Eastwood, N., 101, 107
Edmunds, W. H., 3
Edwards, A. R., 640
Ehnert, G., 538
Eig, M., 143
Eise, K., 621
Ellerhorst, Jr., H., 467
Emrich, P., 582
Englehart, W. H., 462, 480, 541
Ensminger, R., 120
Epel, J. N., 84, 94
Erickson, W. O., 18, 24, 29, 34
Eshleman, L. H., 105, 113, 148, 160
Espenshade, D. J., 201, 221, 232
Evans, D., 320, 498
Fackler, M., 345, 584
Fasco, J. V., 586
Farris, R. D., 531
Farrissey, W. J., 578
Fazekas, E., 428
Fekete, F., 174, 180, 191, 209, 247, 283, 304, 341, 428, 619, 633, 637
Feltzin, J., 537
Ferrarini, J., 499, 500
Ferriday, J. E., 75
Fesco, V., 112
Finsel, J. R., 534

Fintemann, C., 567, 572, 591, 605
Fisher, B. C., 533, 643, 644
Fisk, C., 10
Fletcher, C.W., 342, 373
Flint, A. J., 537
Fock, K., 524
Foley, N. J., 179
Ford, B., 442
Foreman, P., 546
Forger, G., 522
Forsyth, G. E., 301
Foster, J. E., 508, 557
Foster, L. P., 93
Found, M. S., 479
Frankenhoff, E. B., 166, 172
Frick, D. K., 224
Friedman, L. W., 88
Frilette, V., 7
Fry, D. P., 331
Fusco, J. V., 477
Gallagher, R. B., 495
Gandhi, K. S., 406, 430
Gandy, R. C., 424, 431, 585, 626
Gardner, I. J., 444, 477
Gardner, I. L., 586, 608
Garholm, K., 408
Garst, J., 241, 343
Gaylord, M. W., 329
Gentry, R. R., 414, 424, 431, 585
Gerard, J., 571
Gibson, A. G., 622
Golemba, F., 535
Gracrwski, A. S., 513

Grant, I., 537
Gray, Jr., E., 569, 570
Green, D. H., 273
Greening, T., 556
Greenzweig, J. F., 380, 539, 565
Greig, J. W., 31
Griffith, R. M., 418, 576
Grigor, Jr., J. M., 153
Grisch, W. E., 381, 426
Grotke, M., 241, 343
Grove, S. A., 463
Gruenwald, R., 115, 337
Gruskiewicz, M., 536
Guervin, P. R., 273
Guiley, C., 339
Gursky, L., 586
Guzeti, A. J., 38
Hall, A., 290
Halper, W. M., 613
Han, C. D., 627
Handlin, Jr., D. L., 613
Hansen, A. M., 27
Haoriliac, S., 481
Haraguchi, S., 618
Harms, W., 451
Harris, B., 319
Harvey, J. L., 35
Hatanaka, Y., 196
Hellner, G., 436
Hentschel, H., 450
Hersch, P., 122
Hill, H. W., 475
Himebaugh, D. C., 606
Hinchman, G. H., 530
Hindersin, R. R., 285
Hirai, Y., 519
Hirano, H., 249
Hodgson, W. D., 458, 460

Hoefele, A. J., 469
Hoffman, D. F., 130
Hoffman, J.D., 620
Horner, A. H., 529, 555, 567, 579, 591, 605, 614, 643
Horton, A. S., 278
Houlton, P. R., 456
Hovanec, A. J., 76, 78
Howard, F. H., 476, 575, 602
Howatineck, F., 428, 551
Howell, D. M., 204
Howes, W. C., 553, 601
Hughes, G., 165
Huminiski, F. M., 271, 472
Hurst, A. T., 552, 587
Husain, K., 529
Irwin, T. J., 91
Iwami, E., 618
Izzard, K. T., 456, 471, 510
Jackson, R. J., 531
Jackson, R. S., 41, 51, 52, 53, 54, 56, 59, 80
Jacobs, E. F., 501
Jain, M. K., 326
Jerigan, J. W., 175
Jestin, J., 307
Johnson, H. C., 609, 616
Johnson, R. B., 73
Johansson, E., 408
Joneja, S. K., 629
Just, H. A., 480
Jutte, R. B., 267, 432, 598
Kallaur, M., 200, 554, 603
Kamath, V., 495
Kanagendra, M., 642, 644
Kardentz, P. D., 631
Kasische, R. S., 526
Kaull, G. H., 111
Kay, D. J., 279, 300

Keating, J. Z., 379
Kent, G., 461
Kent, J. R., 176
Keown, J. A., 223, 229, 274, 571
Kerr, R. C., 257
Kinnin, J. I., 158
Kirsch, P. A., 596
Ritter, J. R., 528
Kittl, R., 524
Knowles, R. B., 619
Kobayashi, G. S., 417
Koleske, J. U., 372
Korve, E. H., 440
Kressin, D. M., 219
Kroekel, C. H., 155, 242, 258, 264
Kubota, H., 340
Kwan, F. K., 170
Lambert, G. H., 497
Lede, B. L., 602
Lee, D. S., 627
Lee, L. J., 576
Leichtle, I. J., 309, 530
Le May, R. A., 308
Leung, W. D., 620
Levy, M., 387
Liebold, R. W., 473, 561, 636
Lindsay, M. W., 568
Lottig, R., 361
Lottler, C., 450
Loveless, H.S., 558
Lowry, J. R., 221, 232, 433, 472
Luchini, A., 276, 452
Lundberg, J. A., 438
Luzier, W. D., 551
Lynskey, B. M., 635
Maaghul, J., 253, 302, 376

McAssey, E. V., 512
McCabe, M. V., 511
McCarghy, N., 607
McCluskey, J. J., 220, 330, 598
McGarry, F. L., 440, 513
McLaughlin, P. V., 512
McNally, J. S., 250
Mackey, B. A., 547
MaCrae, J. D., 239
MaCrael, T., 347
Madine, T. V., 482
Magrans, J. J., 499, 500
Mandell, J. F., 513
Mandy, F., 321
Marker, L., 442, 576
Marks, J. A., 67
Marple, R. K., 433, 472, 478
Marszewski, C. A., 62
Martinez, V., 637
Massey, F. L., 420
Mate, Z., 588
Mathews, V. E., 431
Mathur, K. K., 422
Mauro, B. E., 299
Maxel, J. M., 166, 172, 188, 369, 419, 426
McKenzie, F. M., 299
McMahan, J. D., 132
Menges, G., 574, 641
Messick, V. B., 631
Methven, J. M., 292, 322, 351, 378, 415
Meyer, R. W., 61, 66, 123, 226, 292, 303, 320, 583
Miller, B. S., 314
Miller, B., 486
Miller, E. K., 166, 172
Miller, E. R., 172, 235

Miller, J. S., 620
Miller, W. P., 620
Milton, A., 307
Moeseller, H. J., 465
Monte, S. J., 425
Moore, B. G., 151
Moore, C. E., 132
Morgan, D. E., 126
Morgan, P., 19
Morris, D. H., 630
Morris, S. J., 558
Morrison, R. S., 293
Mulder, A. C., 87
Muntion, C. B., 12
Myers, F. A., 560, 580
Naitove, M. H., 492
Neiman, T., 361
Nelb, R. G., 506
Nelson, B. W., 12
Nelson, J. L., 565
Nenadal, G. G., 138, 167, 362, 367
Nelb, R. B., 506
Nelson, G., 508
Ness, W. W., 288, 323
Newaz, G. M., 629
Newman, S., 606
Newton, G. P., 471, 510
Nomagucchi, K., 618
Nyo, H., 560
Newton, G. P., 451, 482
Nicklin, N. B., 254
Nussbaum, H. W., 192
O'Hern, T. P., 501
Olsen, N., 434
Owen, M. J., 479
Owens, J. H., 609, 616
Paci, R., 154
Panter, M. R., 564

Parker, F. J., 75, 254, 261, 429, 468
Parr, A. J., 351
Pattison, V. A., 285
Parkyn, B., 5
Pelhan, O. W., 28
Pelton, R., 417
Penfold, R., 404
Pennington, D. W., 248, 464, 502
Peters, J. B., 610
Peterson, P., 483
Pfaff, F., 184
Phillips, D. R., 173
Phillips, T. E., 60
Pickering, Tgt. L., 366
Pinkowski, N. J., 536
Pladek, O., 569
Pluddemann, E. D., 406
Potkanowicz, E. J., 208, 362
Powell, R., 416, 421
Prather, K., 149
Pratt, B. D., 62
Proudfit, C. W., 42
Protrubacz, J. G., 81, 89, 135
Pullman, G. A., 315
Pulman, L. J., 268
Quick, J. R., 557
Quick, W., 511
Rabenold, R., 241
Raichie, K., 149
Raffel, B., 198
Rapp, R. S., 411, 478, 526
Ray, D. J., 479
Reiling, V. G., 14
Reitz, J., 474, 523
Reitz III, J. A., 504
Rheinfrank, G., 82
Richards, W. B., 276

Richter, C. S., 443
Riddell, R. M., 551
Riew, C. K., 327, 422
Ring, E., 240, 264
Ritter, 528
Rizzi, M., 256
Robbins, G., 337
Robertson, F. C., 600
Rockwood, L., 481
Rosato, D. V., 234
Ross, P. M., 370
Ross, R., 558
Ross, R. C., 160
Rowe, E. H., 327, 455, 571
Rutheford, J. L., 603
Sacks, W., 515
Sadler, K. B., 34
Sakaguchi, S., 491
Sanders, B. A., 466
Savage, R. J., 91, 102
Scharfenberger, J. A., 577
Schicktanz, R., 305
Schmeltzer, E., 606
Schmitterger, P. A., 332
Schnell, H., 149
Schnitzler, J., 96
Scholtis, K., 134, 140
Schortenberger, S., 427
Schrantz, J., 624, 625
Schulz, H., 224
Schwarz, E. G., 397
Schwartz, W. T., 272
Schweitzer, W. P., 91
Scott, K. A., 44
Seamark, M. J., 532
Sedlatschek, R. L., 603
Seeman, D. J., 408
Seidl, L. H., 67
Selley, H. E., 266, 276

Seymour, G. M., 260, 283, 293
Shannon, R. F., 24
Shah, N. N., 266
Shank, R. S., 107, 131
Shanowski, H., 401
Shanta, P. L., 61
Sheatsley, R. W., 198, 212
Shenk, W., 320, 377
Sheppard, H. P., 15, 41
Sherer, D. L., 543
Sherman, E. A., 453
Shortall, J. B., 445
Sieglaff, C. L., 250
Signorelli, E., 306
Silva-Nieto, R. S., 505
Silverman, E. M., 550
Sirabian, S., 588
Skogland, G. S., 480
Slayter, G., 9
Simko, T. D., 569
Smith, R. J., 414
Smith, R. L., 258, 280, 358
Smith, S. A., 523
Sonneborn, R. H., 18
South, A., 423, 454
Speelman, D., 584
Spiwak, L., 95
Stanley, A., 483
Stark, C. L., 406
Stavinoha, R. F., 228
Stone, D. H., 180
Stoops, R., 178
Strasser, F., 37
Sturman, D., 118
Sutcliffe, M. R., 44, 77
Sugarman, G., 408
Sundstromn, C. A., 324
Suter, L., 480

Svobada, G. R., 160
Swaneveld, I. A., 430
Tagaki, S., 186
Taylor, E. J., 573
Taylor, R. J., 163
Thomas, J. J., 69, 121
Thomas, J. L., 110
Thomas, D. H., 458
Thomas, R. R., 352, 425
Thompson, E. J., 547
Thompson, P. E. 587
Todd, W. H., 375
Torres, A. F., 45, 90
Tiffin, A. J., 101, 107, 116, 127, 131, 179
Tortolano, F. W., 287, 343, 347
Trimble, S. M., 457
Tropp, F. E., 268
Trudeau, E. G., 538
Tudor, W. E., 227
Uffner, M. W., 274, 323
Vacarri, J. A., 487
Vaill, E. W., 142
Vanderbilt, B. M., 60
Van Gasse, R. L., 441, 493
Vansco-Szmercsamyi, I., 196
Vockel, R. L., 361
Voeks, S. L., 525
Von Tesmar, C., 215
Walker, A. C., 221
Walling, S. J., 217
Walter, O., 325
Walton, J. P., 102, 109, 112, 124, 126, 132, 136, 137, 154, 168, 218, 285, 290, 303
Wang, M., 597
Ward, D. D., 337, 369
Warner, K. N., 269
Water, E. R., 358
Weaver, W. I., 6

Weiner, A. L., 102
Werkkheiser, R. L., 37
White, R. A., 168
White, R. B., 4, 15, 22, 49, 51, 52, 53, 54, 71, 83, 99
Wiehl, F. R., 630
Wildman, G., 617
Wilkinson, W. L., 520
Williams, D. R., 250
Williams, G. L., 371, 395
Williamson, G. A., 622
Williger, E. J., 623
Willis, C. L., 613
Wilson, E. L., 330
Wittman, L., 42
Wolf, G. M., 405
Wood, A. S., 284, 548
Wood, E. L., 508
Wood, R. P., 235
Woodhaua, R. T., 298
Woodrich, R. H., 505
Workman, R. T., 441
Wu, Y. T., 634
Wuh, J., 572, 632
Wright, F. M., 161, 186, 244, 311
Wright, V., 169
Xantos, R. T., 298
Yee, G. Y., 516
Yen, S. C., 630
Yokoo, M., 196
Yonagi, S., 519
Yotsuzuka, M., 249
Younge, K. R., 534
Yovino, J. O., 74, 96
Zaske, O., 529, 555, 567, 572, 579, 591, 632
Ziehm, J. J., 599
Zileg, G. L., 639
Zimmerman, A., 169

Appendix: FRP Glossary

Accelerator: A highly active oxidizing material suspended in a liquid carrier used in conjunction with an initiator to produce internal heat in a liquid resin to affect cure.

Activator: See accelerator.

Air-inhibited Resin: A resin in which surface cure is inhibited or stopped by the presence of air.

B stage: The condition of partially cured resin when it is only partially soluble in monomer or acetone but still is plastic and heat fusible.

Barcol Hardness: A value obtained by use of a Barcol Impressor on the surface of a laminate. Higher values indicate more complete cure.

Bulk Molding Compound (BMC): A mixture of resin, fillers, short glass fibers and ancillary materials required by the end application.

Catalyst: See initiator.

Cavity: The space between matched metal molds in which the laminate or part is formed. Also used to describe the female mold half.

Crazing: Hairline cracks either within or on the surface of a laminate caused by cure-generated stresses.

Cross-linking: The setting up of chemical links between molecular chains.

Cure: The cross-linking or total polymerization of the molecules of a resin. Transformation of the liquid resin to a solid state.

Cure Time: The time required for the liquid resin to reach a cured or fully polymerized state after the initiator has been added.

Delamination: The separation of the laminate along the plane of its reinforcements.

Dimensional Stability: The ability to retain constant shape and size.

Draft: The relief angle used on the sides of a mold to allow part removal.

Exotherm Curve: A chart of temperature versus time during the curing cycle of a resin mix. Peak exotherm is the point of maximum temperature recorded.

Exotherm Heat: Heat given off during a polymerization reaction by the chemical ingredients as they react and cure the resin.

Filament: A single, hairlike thread of glass fiber. The basic part of glass fibers from which yarns are made.

Fillers: See inert fillers.

Finish: The treatment applied to glass fibers to allow the laminating resin to wet them.

Flexural Strength: The ability of a material to withstand bending stresses.

Gel: A partial cure of a resin; a semisolid jellylike state.

Gelcoat: A protective surface coating of clear or pigmented resin.

Gel Time: The time required to change a flowable resin to a nonflowable gel condition.

Hand Lay-up: The oldest and simplest molding technique to which reinforcing materials and catalyzed resin mix are applied on a mold by hand and then compressed with rollers to eliminate entrapped air.

Hardener: See initiator.

Inert Fillers: An inorganic material added to resins to extend volume, improve properties, and reduce costs of the molded part.

Inhibitor: A material that retards polymerization of the monomer and increases the gel time of the mix.

Initiator: A material that produces free radicals to begin the cure of a resin. Frequently referred to as a catalyst.

Laminating: Applying layers of reinforcing materials and resin and eventually bonding those layers together.

Laminate: A material composed of successive layers of reinforcements bonded together with a resin.

Mat: A randomly distributed layer of glass fibers used as a reinforcement in a laminate.

Mold Cavity: See cavity.

Mold Release: A material used to coat the mold to prevent the cured laminate from sticking to the mold.

Monomer: A simple material capable of polymerization.

Orthophthalic Resin: An unsaturated polyester resin in which phthalic anhydride was the starting point.

Polyester Resin: Generally used for unsaturated polyesters.

Polymer: The end product, usually a solid, produced from a monomer.

Porosity: The formation of undesirable clusters of air bubbles in the surface of a laminate.

Pot Life: The length of time that a catalyzed resin mix remains workable.

Preform: A process in which glass mats are formed over a perforated screen shaped like the mold in which the part will be made.

Premix: An admixture of resin, reinforcement, fillers, etc., not in web form, usually prepared by the molder just before use.

Prepreg: A reinforcement saturated with a B-staged resin mix, ready for molding.

Promoter: See activator.

Release Agent: A lubricant, often wax, used to prevent the adhesion of the molded part to the mold.

Resins: A liquid, cross-linkable plastic solution used as a matrix for the reinforcement.

Roving: Continuous strands of glass fibers that are wound together on a spool.

Sheet Molding Compound (SMC): A fiberglass-reinforced thermosetting compound in sheet form.

Shelf Life: The length of time an uncatalyzed resin or resin mix remains workable while stored in a tightly sealed container.

Spray-up: A process in which a chopper gun is used to simultaneously deposit glass fibers and resin mix on a mold.

Styrene Monomer: A water-thin liquid monomer used to dissolve polyester resins and act as a cross-linking agent.

Substrate: Any material that provides a supporting surface for another material.

Surfacing Agent: An oily or waxy material that rises to the surface of a polyester resin during cure.

Tensile Strength: The maximum tensile stress sustained by a specimen before failure in a tensile test.

Thickeners: A material added to a resin mix to chemically thicken the mix by raising its viscosity so that it will not flow as readily.

Thixotropy: The property of being a gel at rest and liquid on agitation.

Undercut: Negative or reverse draft on a mold.

Viscosity: A measure of the resistance to flow of a liquid.

Wet-out: The ability of a resin mix to quickly saturate the reinforcement.

Yarn: A twisted strand or strands of fibers used to form a cloth.

List of Trademarks and Owners

Alperox™ Lucidol Div., Pennwalt Corp.
Altex™ Alpha Corporation
Aropol™ Ashland Oil, Inc.
Audrey™ Tetrahedron Associates
Cadox™ Noury Chemical Corp.
Creel-Paks™ Owens-Corning Fiberglas Corp.
Derakane™ Dow Chemical Co.
Dacron™ E. I. du Pont de Nemours & Co.
Esperox™ U.S. Peroxygen Div.
Flomat™ Fathergil & Harvey
Hetron™ Ashland Oil, Inc.
Hycar™ Goodrich Chemical Co.
ITP™ ICI Americas Inc.
Koplac™ Koppers Co., Inc.
Liladox™ Lucidol Div., Pennwalt Corp.
Luperox™ Lucidol Div., Pennwalt Corp.
Lupersol™ Lucidol Div., Pennwalt Corp.
Mylar™ E. I. du Pont de Nemours & Co.

Percadox™ Noury Chemical Corp.
Pliogrip™ Goodyear Tire & Rubber Co.
Premi-Glas™ Premix Inc.
Silmar™ Silmar Div., Vistron Corp.
Stak-Paks™ PPG Industries
Stypol™ Freeman Chemical Corp.
Teflon™ E. I. du Pont de Nemours & Co.
Triganox™ Noury Chemical Corp.
Twinflow™ Liquid Controls Corp.
Type 30™ Owens-Corning Fiberglas Corp.
USP-245™ U.S. Peroxygen Co.
Vazo™ E. I. du Pont de Nemours & Co.
Vibrinmat™ USS Chemicals
VR-3™ Union Carbide Corp.
WONDERdrill™ International Carbide Corp.
WONDERtwist™ International Carbide Corp.
Zelec™ E. I. du Pont de Nemours & Co.

List of Manufacturers

Allied Moulded Products Inc., P. O. Box 623, Bryan, OH 43506
Alpha Corporation, Collierville, TN 38017
Aluminum Co. of America, 1501 Alcoa Bldg., Pittsburgh, PA 15219
Amoco Chemicals Corp., 200 E. Randolph Dr., Chicago, IL 60601
American Cyanamide Co., Chemical Products Div., One Cyanamide Plaza, Wayne, NJ 07470
Ashland Chemical Co., Div. of Ashland Oil Co., Columbus, OH 43216
Bonnot Co., Sub. of C. L. Gougler Machine Co., 805 Lake St., Kent, OH 44240
I. G. Brenner Company, 32 E. North St., P. O. Box 308, Newark, OH 43055
E. B. Blue Co., 651 Connecticut Ave., South Norwalk, CT 06854
BMC Inc., 3N497 N. 17th St., St. Charles, IL 60174
Brookfield Engineering Laboratories Inc., 240 Cushing St., Stoughton, MA 02072
BYK-Mallinckrodt Chemical Products GmbH, U.S. Marketing Div., 734 Walt Whitman Road, Mellville, NY 11746
Harry T. Campbell Sons Co., Flintcote Corp., Towson, MD 21204
CertainTeed Corp., Fiber Glass Reinforcements, Valley Forge, PA 19482
Creative Pultrusions Inc., Pleasantville Industrial Park, Alum Bank, PA 15521

Day Mixing Co., 4932 Beech St., Cincinnati, OH 45212
Dow Chemical Co., Designed Products Dept., Midland, MI 48640
D-M-E Inc., 29111 Stephenson Hwy., Madison Heights, MI 48071
E. I. du Pont de Nemours & Co., Organic Chemicals Dept., Dyes & Chemicals Div., Wilmington, DE
Eastman Chemical Products Inc., Chemicals Div., Kingsport, TN 37662
Engineering Technology Inc., 145 West 2950 South, Salt Lake City, UT 84115
Englehard Minerals & Chemicals Co., Edison, NJ 08817
Finn & Fram Inc., 13231 Louvre St., Arleta, CA 91331
Freeman Chemical Corp., Port Washington, WI 53074
Farbenfabriken Bayer AG, Werk Uerdingen, 4150 Krefeld-Uerdingen, West Germany
Gale Products Div., Outboard Marine Corp., Galesburg, IL
Georgia Marble Co., Atlanta, GA 30339
Goodrich Chemical Co., Adhesive Products, 500 S. Main St., Akron, OH 44318
Gougler Machine Co., 805 Lake St., Kent, OH 44240
John T. Hepburn, Ltd., 914 Dupont St., Toronto, Canada M6H 1Z2
Hercules Inc., 910 Market St., Wilmington, DE 19899
J. M. Huber Corp., Route 4 Huber, Macon, GA 31298
ICI America, Inc., Wilmington, DE 19897
Indusmin Ltd., Commerce Court West, Toronto, Ont., Canada
International Carbide Corp., 32022 Eighth Ave. S, Roy, WA 98580-9990
Koppers Company, Inc., Koppers Building, Pittsburgh, PA 15219
Lucidol Div., Pennwalt Corp., 1740 Military Rd., Buffalo, NY 14240
Liquid Controls Corp., 7576 Freedom Ave. N.W., North Canton, OH 44720
Martin Hydraulics Inc., 1265 Babbitt Road, Cleveland, OH 44132
NCR Corp., 1700 S. Patterson Blvd., Dayton, OH 45479
The Norac Co., 405 S. Molar Ave., Azuza, CA 91702

NYCO Div., Processed Minerals Inc., Mountain View Road, Willsboro, NY 12996
Noury Chemical Company, Route 78, Burt, NY 14028
Omya, Inc., 61 Main St., Proctor, VT 05765
Outboard Marine Corp., 100 Sea Horse Dr., Waukegan, IL 60085
Owens-Corning Fiberglas Corp., Fiberglas Tower, Toledo, OH 43659
PPG Industries, Fiber Glass Products, Pittsburgh, PA 15272
Pfizer Inc., 235 E. 42nd St., New York, NY 10017
Premix Inc., P. O. Box 281, North Kingsville, OH 44068

Index

Alumina trihydrate, 88
Annotated bibliography, 253
Asbestos, 89

Bibliography, annotated, 253
   Abbreviations, list of, 253
   Author index, 349
Bonded joints, 44
Bosses, 32, 34
Bulk molding compounds
   Description, 2
   Formulations, 127
   History, 2
   Shrinkage, 4
   Specific design details, 63

Calcium carbonate, 79
Carrier films, 109
China clay, 83
Chopped strands, 97
Chrome plating molds, 62
Compression molding, 186
   Steam heating fundamentals, 186
Continuous mixing/molding system, 217
Core pins, 59
Cylinders, mold, 56

Defects, paint, 239
Design
   Mash-off areas, 53
   Shear edges, 51
Dispensers, resin mix, 116
Drilling, 37
Double arm mixers, 166
Dynamic mixers, 123

Ejector sleeves, 59
Extruders, clay-type, 166
Epoxy resins, 75

Fillers, inorganic, 78
   Alumina trihydrate, 89
   Asbestos, 89
   Calcium carbonate, 79
   China clay, 83
   Effect on molding compounds, 80
   Talc, 86
   Wollastonite, 89
Films, carrier, 109
FRP glossary, 357
Fabrication, 36
Fasteners, 39
FRP part design, 19
Flash blowout openings, mold, 56
Flow
   Around corners, 35
   Molding compound, 193
   Thinner to thicker sections, 26

Heating conduits in molds, 55
Heel blocks on molds, 54
High strength compounds, 143

Injection molding, 205
   Effect of glass length, 211
   Process recommendations, 214
   RAM vs. screw, 208
   Runner & gate design, 214
Inserts, 32
Internal release agents, 103
Inorganic fillers, 78

Joint stresses, 43

Knockout pins, 25
Kaolin clay, 83

Leader pins, 55
Low density compounds, 145
Low pressure compounds, 142

Low profile
    Additives, 76
    Discussion, 9

Manufacturers, list of, 365
Master models, 48
Materials
    Comparative properties, 22
Mechanical fasteners, 39
Milling, 38
Mixers
    Blade style, 167
Mold
    Design, 47
    Ejector assembly, 57
    Handling requirements, 60
    Steel selector, 50
    Stops, 56
    Tryout, 62
Molding
    Picture-frame-type parts, 29
Molding compounds
    Electrical type, 13
    Energy absorbing, 16
    Food grade, 15
    Low density, 15
    Low pressure, 15
    Markets, 17
Molds
    Chrome plating, 62
    Duplicating models, 51
    Heel blocks, 54
    Internal heating conduits, 55
    Leader pins, 55
    Mash-off designs, 53
    Master models, 48
    Mold steel selector, 50
    Shear edge designs, 51
    Surface finish, 61
    Types, 12

Organic peroxides, 98

Painting
    Automotive primers, 230
    Automotive sealers, 231
    Automotive topcoats, 233
    Application equipment, 237
    Defects, 239
    Defect definitions, 240
    Paint components, 222
    Paint requirements, 227
    Paint types, 264
    Surface preparation, 228
Part design
    Blind pockets, 24
    Draft, 24
    Flash location, 21
    Parting lines, 23
    Stress-strain curves, various materials, 20
Pigments, 106
Polyester resins
    Ingredients, 66
    Types, 68
Premix
    Advantages, 1
    Disadvantages, 1
Preparing BMC charges, 195
Preparing resin mixes, 115
Preparing SMC charges, 196
Punching, 38

Raw materials, 65
Reinforcing mats, 96
Ribs
    Appearance parts, 32
    Nonappearance parts, 31
Routing, 37
Roving choppers, 156
Roving, fiberglass, 91
    Terms, definition of, 92

Sawing, 38

Sheet molding compounds
    Description, 5
    Effect of glass content, 6
    Formulations, 137
    History, 5
    Impregnator design, 156
    In-process testing, 123
    Low profile discussion, 9
    Test procedures, 125
    Thickening mechanism, 11
Static mixers, 122
Surface finish, mold, 61

Talc, 86

Tapping, 39
Testing molding compounds
    Physical properties, 191
    Reactivity, 189
    Spiral flow, 191
    Viscosity increase of mixes, 187
Thermoplastic additives, 76
Thickeners, chemical, 110
Thickening mechanism, 6
Toughener additives, 111
Trademarks, list of, 363
Troubleshooting guide
    BMC injection molding, 217
    SMC molding, 199
    SMC processing, 118

DISCARDED

JUN 2 6 2025

DATE DUE
NOV 28-94

**ASHEVILLE-BUNCOMBE TECHNICAL COLLEGE
LEARNING RESOURCES CENTER
340 VICTORIA ROAD
ASHEVILLE, NC 28801**